Introduction to Synthetic Aperture Radar (SAR) Polarimetry

by

Wolfgang-Martin Boerner

Sensing & Navigation Laboratory

Wexford Press
2008

Table of Contents

BASICS OF SAR POLARIMETRY I

Wolfgang-Martin Boerner

Basics of Radar Polarimetry

Abstract A comprehensive overview of the basic principles of radar polarimetry is presented. The relevant fundamental field equations are first provided. The importance of the propagation and scattering behavior in various frequency bands, the electrodynamic foundations such as Maxwell's equations, the Helmholtz vector wave equation and especially the fundamental laws of polarization will first be introduced: The fundamental terms which represent the polarization state will be introduced, defined and explained. Main points of view are the polarization Ellipse, the polarization ratio, the Stokes Parameter and the Stokes and Jones vector formalisms as well as its presentation on the Poincaré sphere and on relevant map projections. The Polarization Fork descriptor and the associated van Zyl polarimetric power density and Agrawal polarimetric phase correlation signatures will be introduced also in order to make understandable the polarization state formulations of electromagnetic waves in the frequency domain. The polarization state of electromagnetic waves under scattering conditions i.e. in the radar case will be described by matrix formalisms. Each scatterer is a polarization transformer; under normal conditions the transformation from the transmitted wave vector to the received wave vector is linear and this behavior, principally, will be described by a matrix called scattering matrix. This matrix contains all the information about the scattering process and the scatterer itself. The different relevant matrices, the respective terms like Jones Matrix, S-matrix, Müller M-matrix, Kennaugh K-matrix, etc. and its interconnections will be defined and described together with change of polarization bases transformation operators, where upon the optimal (Characteristic) polarization states are determined for the coherent and partially coherent cases, respectively. The lecture is concluded with a set of simple examples.

1. Introduction: A Review of Polarimetry

Radar Polarimetry (*Polar*: polarization, *Metry*: measure) is the science of acquiring, processing and analyzing the polarization state of an electromagnetic field. Radar polarimetry is concerned with the utilization of polarimetry in radar applications as reviewed most recently in Boerner [1] where a host of pertinent references are provided. Although polarimetry has a long history which reaches back to the 18th century, the earliest work that is related to radar dates back to the 1940s. In 1945 G.W. Sinclair introduced the concept of the scattering matrix as a descriptor of the radar cross section of a coherent scatterer [2], [3]. In the late 1940s and the early 1950s major pioneering work was carried out by E.M. Kennaugh [4, 5]. He formulated a backscatter theory based on the eigenpolarizations of the scattering matrix introducing the concept of optimal polarizations by implementing the concurrent work of G.A. Deschamps, H. Mueller, and C. Jones. Work continued after Kennaugh, but only a few notable contributions, as those of G.A. Deschamps 1951 [6], C.D. Graves 1956 [7], and J.R. Copeland 1960 [8], were made until Huynen's studies in 1970s. The beginning of a new age was the treatment presented by J.R. Huynen in his doctoral thesis of 1970 [9], where he exploited Kennaugh's optimal polarization concept [5] and formulated his approach to target radar phenomenology. With this thesis, a renewed interest for radar polarimetry was raised. However, the full potential of radar polarimetry was never fully realized until the early 1980s, due in no small parts to the advanced radar device technology [10, 11]. Technological problems led to a series of negative conclusions in the 1960s and 1970s about the practical use of radar systems with polarimetric capability [12]. Among the major contributions of the 1970s and 1980s are those of W-M Boerner [13, 14, 15] who pointed out the importance of polarization first in addressing vector electromagnetic inverse scattering [13]. He initiated a critical analysis of Kennaugh's and Huynen's work and extended Kennaugh's optimal polarization theory [16]. He has been influential in causing the radar community to recognize the need of polarimetry in remote sensing applications. A detailed overview on the history of polarimetry can be found in [13, 14, 15], while a historical review of polarimetric radar technology is also given in [13, 17, 18].

Boerner, W.-M. (2007) Basics of SAR Polarimetry I. In *Radar Polarimetry and Interferometry* (pp. 3-1 – 3-40). Educational Notes RTO-EN-SET-081bis, Paper 3. Neuilly-sur-Seine, France: RTO. Available from: http://www.rto.nato.int/abstracts.asp.

Polarimetry deals with the full vector nature of polarized (vector) electromagnetic waves throughout the frequency spectrum from Ultra-Low-Frequencies (ULF) to above the Far-Ultra-Violet (FUV) [19, 20]. Whenever there are abrupt or gradual changes in the index of refraction (or permittivity, magnetic permeability, and conductivity), the polarization state of a narrow band (single-frequency) wave is transformed, and the electromagnetic "vector wave" is re-polarized. When the wave passes through a medium of changing index of refraction, or when it strikes an object such as a radar target and/or a scattering surface and it is reflected; then, characteristic information about the reflectivity, shape and orientation of the reflecting body can be obtained by implementing '*polarization control*' [10, 11]. The complex direction of the electric field vector, in general describing an ellipse, in a plane transverse to propagation, plays an essential role in the interaction of electromagnetic '*vector waves*' with material bodies, and the propagation medium [21, 22, 13, 14, 16]. Whereas, this polarization transformation behavior, expressed in terms of the "polarization ellipse" is named "*Ellipsometry*" in Optical Sensing and Imaging [21, 23], it is denoted as "*Polarimetry*" in Radar, Lidar/Ladar and SAR Sensing and Imaging [12, 14, 15, 19] - using the ancient Greek meaning of "*measuring orientation and object shape*". Thus, *ellipsometry* and *polarimetry* are concerned with the control of the coherent polarization properties of the optical and radio waves, respectively [21, 19]. With the advent of optical and radar polarization phase control devices, *ellipsometry* advanced rapidly during the Forties (Mueller and Land [24, 21]) with the associated development of mathematical *ellipsometry*, i.e., the introduction of '*the 2 x 2 coherent Jones forward scattering (propagation) and the associated 4 x 4 average power density Mueller (Stokes) propagation matrices*' [21]; and *polarimetry* developed independently in the late Forties with the introduction of dual polarized antenna technology (Sinclair, Kennaugh, et al. [2, 3, 4, 5]), and the subsequent formulation of '*the 2 x 2 coherent Sinclair radar back-scattering matrix and the associated 4 x 4 Kennaugh radar back-scattering power density matrix*', as summarized in detail in Boerner et al. [19, 25]. Since then, *ellipsometry* and *polarimetry* have enjoyed steep advances; and, a mathematically coherent polarization matrix formalism is in the process of being introduced for which the lexicographic covariance matrix presentations [26, 27] of signal estimation theory play an equally important role in *ellipsometry* as well as *polarimetry* [19]. Based on Kennaugh's original pioneering work on discovering the properties of the "Spinorial Polarization Fork" concept [4, 5], Huynen [9] developed a "*Phenomenological Approach to Radar Polarimetry*", which had a subtle impact on the steady advancement of *polarimetry* [13, 14, 15] as well as *ellipsometry* by developing the "*orthogonal (group theoretic) target scattering matrix decomposition*" [28, 29, 30] and by extending the characteristic optimal polarization state concept of Kennaugh [31, 4, 5], which lead to the renaming of the spinorial polarization fork concept to the so called '*Huynen Polarization Fork*' in '*Radar Polarimetry*' [31]. Here, we emphasize that for treating the general bistatic (asymmetric) scattering matrix case, a more general formulation of fundamental *Ellipsometry* and *Polarimetry* in terms of a spinorial group-theoretic approach is strictly required, which was first explored by Kennaugh but not further pursued by him due to the lack of pertinent mathematical formulations [32, 33].

In *ellipsometry*, the Jones and Mueller matrix decompositions rely on a product decomposition of relevant optical measurement/transformation quantities such as diattenuation, retardence, depolarization, birefringence, etc., [34, 35, 23, 28, 29] measured in a '*chain matrix arrangement, i.e., multiplicatively placing one optical decomposition device after the other*'. In *polarimetry*, the Sinclair, the Kennaugh, as well as the covariance matrix decompositions [29] are based on a group-theoretic series expansion in terms of the principal orthogonal radar calibration targets such as the sphere or flat plate, the linear dipole and/or circular helical scatterers, the dihedral and trihedral corner reflectors, and so on - - observed in a linearly superimposed aggregate measurement arrangement [36, 37]; leading to various canonical target feature mappings [38] and sorting as well as scatter-characteristic decomposition theories [39, 27, 40]. In addition, polarization-dependent speckle and noise reduction play an important role in both *ellipsometry* and *polarimetry, which in radar polarimetry were first pursued with rigor by J-S. Lee* [41, 42, 43, 44]. The implementation of all of these novel methods will fail unless one is given fully calibrated scattering matrix information, which applies to each element of the Jones and Sinclair matrices.

It is here noted that it has become common usage to replace "ellipsometry" by "optical polarimetry" and expand "polarimetry" to "radar polarimetry" in order to avoid confusion [45, 18], a nomenclature adopted in the remainder of this paper.

Very remarkable improvements beyond classical "non-polarimetric" radar target detection, recognition and discrimination, and identification were made especially with the introduction of the covariance matrix optimization procedures of Tragl [46], Novak et al. [47 - 51], Lüneburg [52 - 55], Cloude [56], and of Cloude and Pottier [27]. Special attention must be placed on the *'Cloude-Pottier Polarimetric Entropy H, Anisotropy A, Feature-Angle ($\overline{\alpha}$) parametric decomposition'* [57] because it allows for unsupervised target feature interpretation [57, 58]. Using the various fully polarimetric (scattering matrix) target feature syntheses [59], polarization contrast optimization, [60, 61] and polarimetric entropy/anisotropy classifiers, very considerable progress was made in interpreting and analyzing POL-SAR image features [62, 57, 63, 64, 65, 66]. This includes the reconstruction of *'Digital Elevation Maps (DEMs)'* directly from *'POL-SAR Covariance-Matrix Image Data Takes'* [67 - 69] next to the familiar method of DEM reconstruction from IN-SAR Image data takes [70, 71, 72]. In all of these techniques well calibrated scattering matrix data takes are becoming an essential pre-requisite without which little can be achieved [18, 19, 45, 73]. In most cases the *'multi-look-compressed SAR Image data take MLC- formatting'* suffices also for completely polarized SAR image algorithm implementation [74]. However, in the sub-aperture polarimetric studies, in *'Polarimetric SAR Image Data Take Calibration'*, and in *'POL-IN-SAR Imaging'*, the *'SLC (Single Look Complex) SAR Image Data Take Formatting'* becomes an absolute must [19, 1]. Of course, for SLC-formatted Image data, in particular, various speckle-filtering methods must be applied always. Implementation of the *'Lee Filter'* – explored first by Jong-Sen Lee - for speckle reduction in polarimetric SAR image reconstruction, and of the *' Polarimetric Lee-Wishart distribution'* for improving image feature characterization have further contributed toward enhancing the interpretation and display of high quality SAR Imagery [41 – 44, 75].

2. The Electromagnetic Vector Wave and Polarization Descriptors

The fundamental relations of radar polarimetry are obtained directly from Maxwell's equations [86, 34], where for the source-free isotropic, homogeneous, free space propagation space, and assuming IEEE standard [102] time-dependence $\exp(+j\omega t)$, the electric $\mathbf{E}$ and magnetic $\mathbf{H}$ fields satisfy with μ being the free space permeability and ε the free space permittivity

$$\nabla \mathrm{x}\mathbf{E}(\mathbf{r}) = -j\omega\mu\mathbf{H}(\mathbf{r}), \qquad \nabla \mathrm{x}\mathbf{H}(\mathbf{r}) = j\omega\varepsilon\, \mathbf{E}(\mathbf{r}) \tag{2.1}$$

which for the time-invariant case, result in

$$(\nabla + k^2)\mathbf{E} = 0, \qquad \mathbf{E}(\mathbf{r}) = E_0 \frac{\exp(-jkr)}{r}, \qquad \mathbf{H}(\mathbf{r}) = H_0 \frac{\exp(-jkr)}{r} \tag{2.2}$$

for an outgoing spherical wave with propagation constant $k = \omega\ (\varepsilon\, \mu)^{1/2}$ and $c = (\varepsilon\, \mu)^{-1/2}$ being the free space velocity of electromagnetic waves

No further details are presented here, and we refer to Stratton [86], Born and Wolf [34] and Mott [76] for full presentations.

2.1 Polarization Vector and Complex Polarization Ratio

With the use of the standard spherical coordinate system $(r,\theta,\phi\ ;\ \hat{\mathrm{u}}_r, \hat{\mathrm{u}}_\theta, \hat{\mathrm{u}}_\phi)$ with r,θ,ϕ denoting the radial, polar, azimuthal coordinates, and $\hat{\mathrm{u}}_r, \hat{\mathrm{u}}_\theta, \hat{\mathrm{u}}_\phi$ the corresponding unit vectors, respectively; the outward travelling wave is expressed as

$$\mathbf{E} = \hat{\mathrm{u}}_\theta\, E_\theta + \hat{\mathrm{u}}_\phi E_\phi \quad \mathbf{H} = \hat{\mathrm{u}}_\theta\, H_\theta + \hat{\mathrm{u}}_\phi\, H_\phi\ , \ \mathbf{P} = \frac{\hat{\mathrm{u}}_r}{2}\left|\mathbf{E}\times\mathbf{H}^*\right| = \frac{\hat{\mathrm{u}}_r\,|\mathrm{E}|^2}{2Z_0}\ ,\ Z_0 = \left(\frac{\mu_0}{\varepsilon_0}\right)^{1/2} = 120\pi\ [\Omega] \tag{2.3}$$

with $\mathbf{P}$ denoting the Poynting power density vector, and Z_0 being the intrinsic impedance of the medium (here vacuum). Far from the antenna in the far field region [86, 76], the radial waves of (2.2) take on plane wave characteristics, and assuming the wave to travel in positive z-direction of a right-handed Cartesian coordinate system (x, y, z), the electric field $\mathbf{E}$, denoting the polarization vector, may be rewritten as

$$\mathbf{E} = \hat{\mathrm{u}}_x E_x + \hat{\mathrm{u}}_y E_y = |E_x| \exp(j\phi_x)\{\hat{\mathrm{u}}_x + \hat{\mathrm{u}}_y \left|\frac{E_y}{E_x}\right| \exp(j\phi)\} \tag{2.4}$$

with $|E_x|, |E_y|$ being the amplitudes, ϕ_x, ϕ_y the phases, $\phi = \phi_y - \phi_x$ the relative phase; $|E_x / E_y| = \tan\alpha$ with ϕ_x, ϕ_y, α and ϕ defining the Deschamps parameters [6, 103]. Using these definitions, the '*normalized complex polarization vector* $\mathbf{p}$' and the '*complex polarization ratio* ρ' can be defined as

$$\mathbf{p} = \frac{\mathbf{E}}{|\mathbf{E}|} = \frac{\hat{\mathrm{u}}_x E_x + \hat{\mathrm{u}}_y E_y}{|\mathbf{E}|} = \frac{E_x}{|\mathbf{E}|}\left(\hat{\mathrm{u}}_x + \rho\, \hat{\mathrm{u}}_y\right) \tag{2.5}$$

with $|\mathbf{E}|^2 = \mathbf{E}\cdot\mathbf{E}^* = E_x^{\ 2} + E_y^{\ 2}$ and $|\mathbf{E}| = E$ defines the wave amplitude, and ρ is given by

$$\rho = \frac{E_y}{E_x} = \left|\frac{E_y}{E_x}\right| \exp(j\phi), \qquad \phi = \phi_y - \phi_x \tag{2.6}$$

2.2 The Polarization Ellipse and its Parameters

The tip of the real time-varying vector $\mathbf{E}$, or $\mathbf{p}$, traces an ellipse for general phase difference ϕ, where we distinguish between right-handed (clockwise) and left-handed (counter-clockwise) when viewed by the observer in direction of the travelling wave [76, 19], as shown in Fig. 2.1 for the commonly used horizontal H (by replacing x) and vertical V (by replacing y) polarization states.

There exist unique relations between the alternate representations, as defined in Fig. 2.1 and Fig. 2.2 with the definition of the orientation ψ and ellipticity χ angles expressed, respectively, as

$$\alpha = |\rho| = \frac{|E_y|}{|E_x|},\ 0 \le \alpha \le \pi/2 \quad \text{and} \quad \tan 2\psi = \tan(2\alpha)\cos\phi \qquad -\pi/2 \le \psi \le +\pi/2 \tag{2.7}$$

$$\tan\chi = \pm\textit{minor axis/major axis}, \qquad \sin 2\chi = \sin 2\alpha \sin\phi\,, \qquad -\pi/4 \le \chi \le \pi/4 \tag{2.8}$$

where the $+$ and $-$ signs are for left- and right-handed polarizations respectively.

For a pair of orthogonal polarizations $\mathbf{p}_1$ and $\mathbf{p}_2 = \mathbf{p}_{1_\perp}$

$$\mathbf{p}_1 \cdot \mathbf{p}_2^{\,*} = 0 \qquad \rho_2 = \rho_{1\perp} = -1/\rho_1^*\,, \qquad \psi_1 = \psi_2 + \frac{\pi}{2} \qquad \chi_1 = -\chi_2 \tag{2.9}$$

In addition, the following useful transformation relations exist:

$$\rho = \frac{\cos 2\chi \sin 2\psi + j \sin 2\chi}{1 + \cos 2\chi \cos 2\psi} = \tan\alpha \exp(j\phi) \tag{2.10}$$

where (α,ϕ) and (ψ,χ) are related by the following equations:

$$\cos 2\alpha = \cos 2\psi \cos 2\chi, \quad \tan\phi = \tan 2\chi / \sin 2\psi \tag{2.11}$$

and inversely

$$\psi = \frac{1}{2}\arctan\left(\frac{2\,\mathrm{Re}\{\rho\}}{1-\rho\rho^*}\right) + \pi \quad \textbf{... mod}(\pi) \qquad \chi = \frac{1}{2}\arcsin\left(\frac{2\,\mathrm{Im}\{\rho\}}{1-\rho\rho^*}\right) \tag{2.12}$$

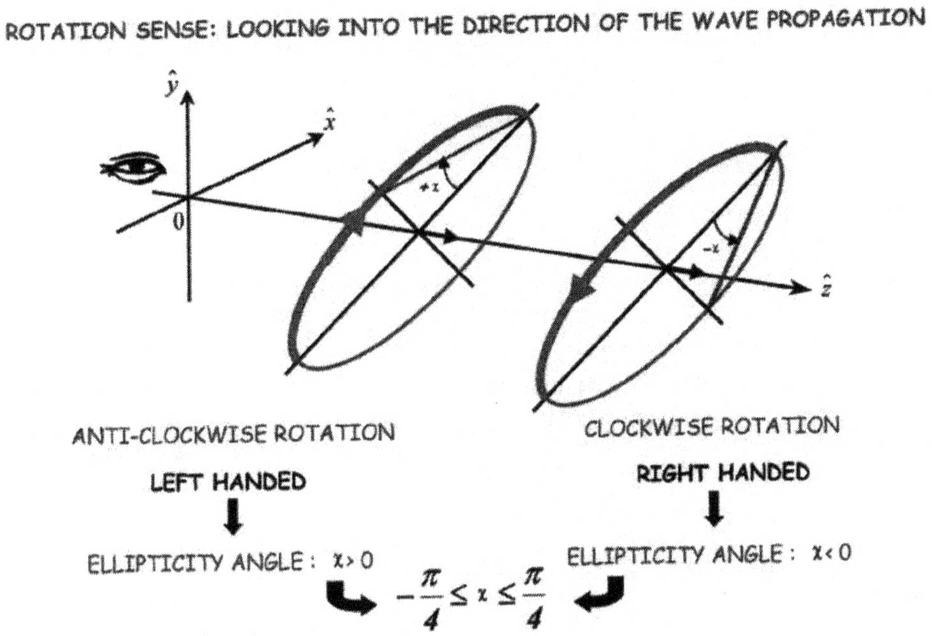

(a) Rotation Sense (Courtesy of Prof. E. Pottier)

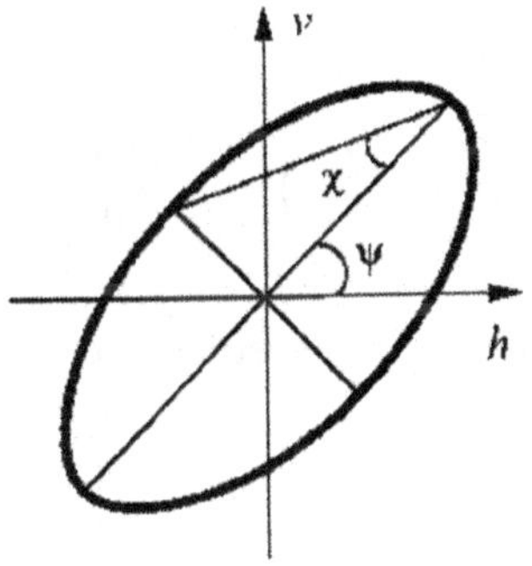

(b) Orientation ψ and Ellipticity χ Angles.

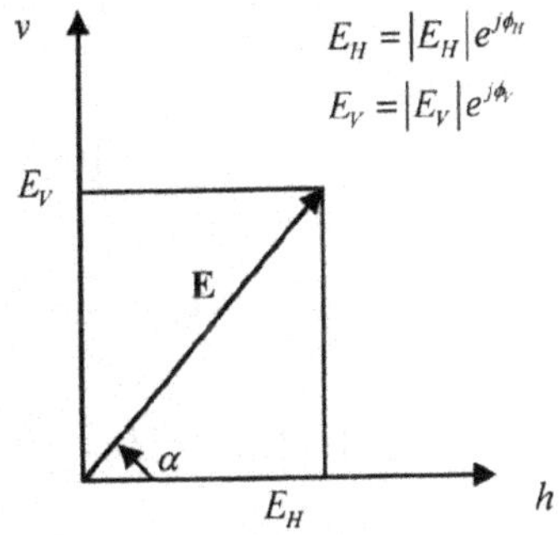

(c) Electric Field Vector.

Fig. 2.1 Polarization Ellipse.

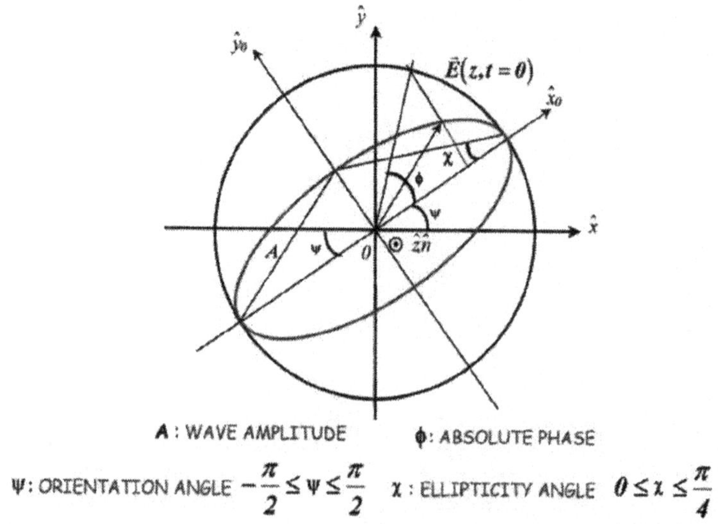

Fig. 2.2 Polarization Ellipse Relations (Courtesy of Prof. E. Pottier)

Another useful formulation of the polarization vector $\mathbf{p}$ was introduced by Huynen in terms of the parametric formulation [9, 104], derived from group-theoretic considerations based on the Pauli SU(2) matrix set $\psi_P\{[\sigma_i]\ ,\ i=0,1,2,3\}$ as further pursued by Pottier [105], where according to (2.10) and (2.11), for $\psi=0$, and then rotating this ellipse by ψ.

$$\mathbf{p}(|\mathbf{E}|,\phi,\psi,\chi)=|\mathbf{E}|\exp(j\phi)\begin{bmatrix}\cos\psi & -\sin\psi\\ \sin\psi & \cos\psi\end{bmatrix}\begin{bmatrix}\cos\chi\\ -j\sin\chi\end{bmatrix} \tag{2.13}$$

which will be utilized later on; and $\psi_P\{[\sigma_i]\ ,\ i=0,1,2,3\}$ is defined in terms of the classical unitary Pauli matrices $[\sigma_i]$ as

$$[\sigma_0]=\begin{bmatrix}1 & 0\\ 0 & 1\end{bmatrix},\quad [\sigma_1]=\begin{bmatrix}1 & 0\\ 0 & -1\end{bmatrix},\quad [\sigma_2]=\begin{bmatrix}0 & 1\\ 1 & 0\end{bmatrix},\quad [\sigma_3]=\begin{bmatrix}0 & -j\\ j & 0\end{bmatrix} \tag{2.14}$$

where the $[\sigma_i]$ matrices satisfy the unitarity condition as well as commutation properties given by

$$[\sigma_i]^{-1}=[\sigma_i]^{T*},\quad |Det\{[\sigma_i]\}|=1,\quad [\sigma_i][\sigma_j]=-[\sigma_j][\sigma_i],\quad [\sigma_i][\sigma_i]=[\sigma_0] \tag{2.15}$$

satisfying the ordinary matrix product relations.

2.3 The Jones Vector and Changes of Polarization Bases

If instead of the basis {x y} or {H V}, we introduce an alternative presentation {m n} as a linear combination of two arbitrary orthonormal polarization states $\mathbf{E}_m$ and $\mathbf{E}_n$ for which

$$\mathbf{E}=\ \hat{\mathbf{u}}_m E_m+\hat{\mathbf{u}}_n E_n \tag{2.16}$$

and the standard basis vectors are in general, orthonormal, i.e.

$$\hat{u}_m \cdot \hat{u}_n^{\dagger} = 0, \quad \hat{u}_m \cdot \hat{u}_m^{\dagger} = \hat{u}_n \cdot \hat{u}_n^{\dagger} = 1 \tag{2.17}$$

with $\dagger$ denoting the hermitian adjoint operator [21, 52, 53]; and the Jones vector $\mathbf{E}_{mn}$ may be defined as

$$\mathbf{E}_{mn} = \begin{bmatrix} E_m \\ E_n \end{bmatrix} = \begin{bmatrix} |E_m| \exp j\phi_m \\ |E_n| \exp j\phi_n \end{bmatrix} = E_m \begin{bmatrix} 1 \\ \rho \end{bmatrix} = \frac{|\mathbf{E}| \exp(j\phi_m)}{\sqrt{1+\rho\rho^*}} \begin{bmatrix} 1 \\ \rho \end{bmatrix} = |\mathbf{E}| \exp(j\phi_m) \begin{bmatrix} \cos\alpha \\ \sin\alpha \exp(j\phi) \end{bmatrix} \tag{2.18}$$

with $\tan\alpha = | E_n / E_m |$ and $\phi = \phi_n - \phi_m$. This states that the Jones vector possesses, in general, four degrees of freedom. The Jones vector descriptions for characteristic polarization states are provided in Fig. 2.3.

$$\mathbf{E}_{mn} = \mathbf{E}(m,n) = \begin{bmatrix} E_m \\ E_n \end{bmatrix} \qquad \mathbf{E}_{ij} = \mathbf{E}(i,j) = \begin{bmatrix} E_i \\ E_j \end{bmatrix} \quad \text{and} \quad \mathbf{E}_{AB} = \mathbf{E}(A,B) = \begin{bmatrix} E_A \\ E_B \end{bmatrix} \tag{2.20}$$

The unique transformation from the $\{\hat{\mathbf{u}}_m \ \hat{\mathbf{u}}_n\}$ to the arbitrary $\{\hat{\mathbf{u}}_i \ \hat{\mathbf{u}}_j\}$ or $\{\hat{\mathbf{u}}_A \ \hat{\mathbf{u}}_B\}$ bases is sought which is a linear transformation in the two-dimensional complex space so that

$$\mathbf{E}_{ij} = [U_2]\mathbf{E}_{mn} \quad or \quad \mathbf{E}(i,j) = [U_2]\mathbf{E}(m,n) \quad \text{with} \quad [U_2]\,[U_2]^{\dagger} = [I_2] \tag{2.21}$$

satisfying wave energy conservation with $[I_2]$ being the 2x2 identity matrix, and we may choose, as shown in [81],

$$\hat{\mathbf{u}}_i = \frac{\exp(j\phi_i)}{\sqrt{1+\rho\rho^*}} \begin{bmatrix} 1 \\ \rho \end{bmatrix} \quad \text{and} \quad \hat{\mathbf{u}}_j = \hat{\mathbf{u}}_{i_\perp} = \frac{\exp(j\phi_i)}{\sqrt{1+\rho\rho^*}} \begin{bmatrix} 1 \\ -\rho^{*-1} \end{bmatrix} = \frac{\exp(j\phi_i)}{\sqrt{1+\rho\rho^*}} \begin{bmatrix} -\rho^* \\ 1 \end{bmatrix} \tag{2.22}$$

with $\phi_j' = \phi_j + \phi + \pi$ so that

$$[U_2] = \frac{1}{\sqrt{1+\rho\rho^*}} \begin{bmatrix} \exp(j\phi_i) & -\rho^* \exp(j\phi_j) \\ \rho \exp(j\phi_i) & \exp(j\phi_j) \end{bmatrix} \tag{2.23}$$

yielding $Det\{[U_2]\} = \exp\{j(\phi_i + \phi_j')\}$ with $\phi_i + \phi_j' = 0$

Since any monochromatic plane wave can be expressed as a linear combination of two orthonormal linear polarization states, defining the reference polarization basis, there exist an infinite number of such bases {i j} or {A B} for which

$$\mathbf{E} = \hat{u}_m E_m + \hat{u}_n E_n = \hat{u}_i E_i + \hat{u}_j E_j = \hat{u}_A E_A + \hat{u}_B E_B \tag{2.19}$$

with corresponding Jones vectors presented in two alternate, most commonly used notations

HORIZONTAL

$$\underline{H} = \begin{bmatrix} 1 \\ 0 \end{bmatrix}$$

$$\psi = 0$$

$$\chi = 0$$

VERTICAL

$$\underline{V} = \begin{bmatrix} 0 \\ 1 \end{bmatrix}$$

$$\psi = \frac{\pi}{2}$$

$$\chi = 0$$

LINEAR

$$\underline{L} = \begin{bmatrix} cos\,\theta \\ sin\,\theta \end{bmatrix}$$

$$\psi = \theta$$

$$\chi = 0$$

ORTHOGONAL LINEAR

$$\underline{L}_\perp = \begin{bmatrix} -sin\,\theta \\ cos\,\theta \end{bmatrix}$$

$$\psi = \theta + \frac{\pi}{2}$$

$$\chi = 0$$

LEFT CIRCULAR

$$\underline{LC} = \frac{1}{\sqrt{2}} \begin{bmatrix} 1 \\ j \end{bmatrix}$$

$$-\frac{\pi}{2} \le \psi \le +\frac{\pi}{2}$$

$$\chi = +\frac{\pi}{4}$$

RIGHT CIRCULAR

$$\underline{RC} = \frac{1}{\sqrt{2}} \begin{bmatrix} 1 \\ -j \end{bmatrix}$$

$$-\frac{\pi}{2} \le \psi \le +\frac{\pi}{2}$$

$$\chi = -\frac{\pi}{4}$$

ELLIPTICAL

$$\underline{E} = \begin{bmatrix} E_x \\ E_y \end{bmatrix}$$

$$\chi = \theta$$

$$0 \le \psi \le +\frac{\pi}{4}$$

ORTHOGONAL ELLIPTICAL

$$\underline{E}_\perp = \begin{bmatrix} E'_x \\ E'_y \end{bmatrix}$$

$$\chi = \theta + \frac{\pi}{2}$$

$$-\frac{\pi}{4} \le \psi \le 0$$

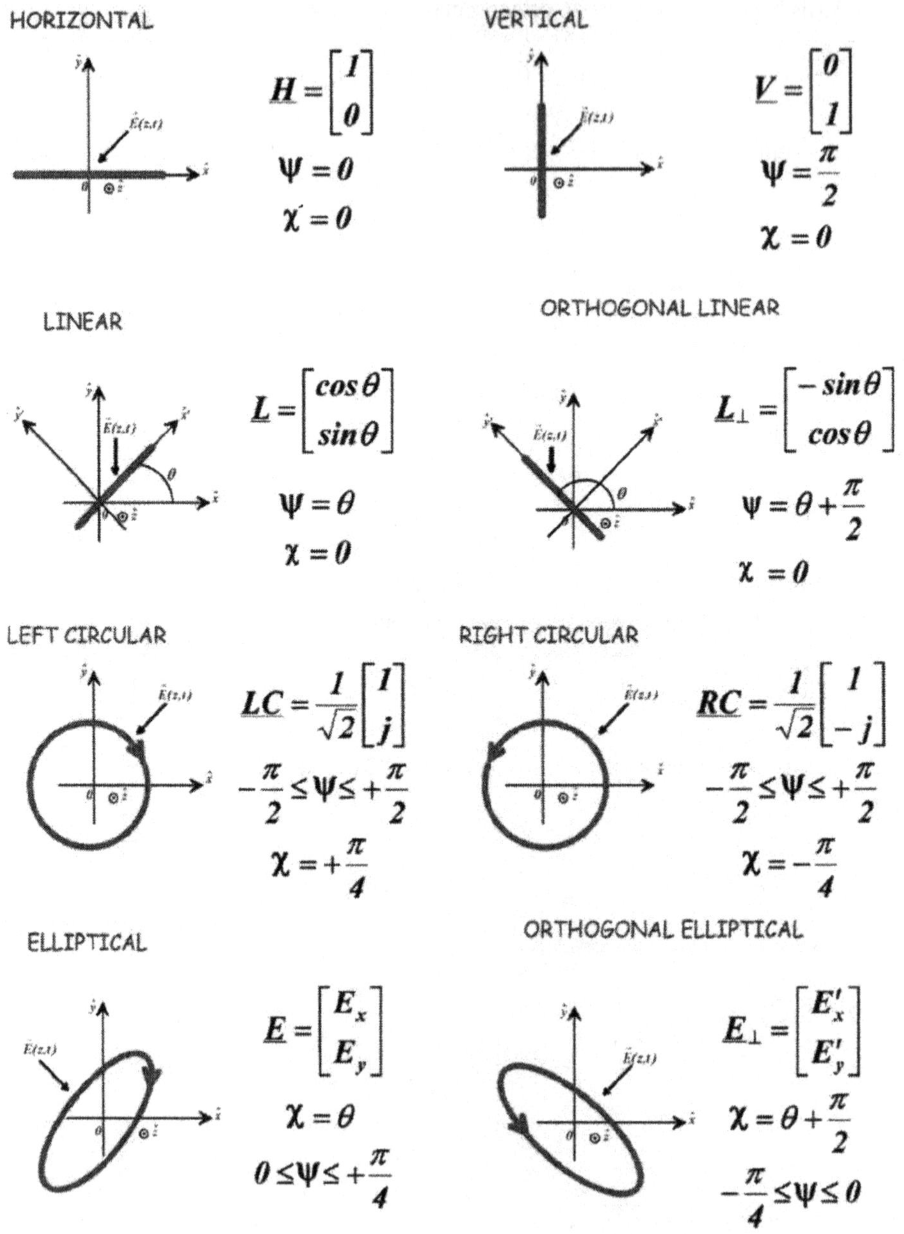

Fig. 2.3 Jones Vector Descriptions for Characteristic Polarization States with direction of propagation out of the page (Courtesy of Prof. E. Pottier)

Since $[U_2]$ is a special unitary 2x2 complex matrix with unit determinant, implying that (i) the amplitude of the wave remains independent of the change of the polarization basis, and that (ii) the phase of the (absolute) wave may be consistently defined as the polarization basis is changed, we finally obtain,

$$[U_2] = \frac{1}{\sqrt{1 + \rho\rho^*}} \begin{bmatrix} 1 & -\rho^* \\ \rho & 1 \end{bmatrix} \begin{bmatrix} \exp(j\phi_i) & 0 \\ 0 & \exp(j\phi_i) \end{bmatrix} \tag{2.24}$$

possessing three degrees of freedom similar to the normalized Jones vector formulation, but in most cases the phase reference is taken as $\phi_i = 0$ which may not be so in polarimetric interferometry [96]. For further details on the group-theoretic representations of the proper transformation relations see the formulations derived by Pottier in [106].

2.4 Complex Polarization Ratio in Different Polarization Bases

Any wave can be resolved into two orthogonal components (linearly, circularly, or elliptically polarized) in the plane transverse to the direction of propagation. For an arbitrary polarization basis {A B} with unit vectors $\hat{a}$ and $\hat{b}$, one may define the polarization state

$$\mathbf{E}(AB) = E_A \hat{a} + E_B \hat{b} \tag{2.25}$$

where the two components E_A and E_B are complex numbers. The polarization ratio ρ_{AB} in an arbitrary basis {A B} is also a complex number, and it may be defined as

$$\rho_{AB} = \frac{E_B}{E_A} = \frac{|E_B|}{|E_A|}\exp\{j(\phi_B - \phi_A)\} = |\rho_{AB}|\exp\{j\phi_{AB}\} \tag{2.26}$$

where $|\rho_{AB}|$ is the ratio of magnitude of two orthogonal components of the field $|E_A|$ and $|E_B|$ and ϕ_{AB} is the phase difference between E_A and E_B. The complex polarization ratio ρ_{AB} depends on the polarization basis {A B} and can be used to specify the polarization of an electromagnetic wave

$$\mathbf{E}(AB) = \begin{bmatrix} E_A \\ E_A \end{bmatrix} = |E_A|\exp\{j\phi_A\}\begin{bmatrix} 1 \\ \rho_{AB} \end{bmatrix} = |E_A|\exp\{j\phi_A\}\frac{\sqrt{1+\dfrac{E_B E_B^*}{E_A E_A^*}}}{\sqrt{1+\dfrac{E_B E_B^*}{E_A E_A^*}}}\begin{bmatrix} 1 \\ \rho_{AB} \end{bmatrix} \tag{2.27}$$

$$= |\mathbf{E}|\exp\{j\phi_A\}\frac{1}{\sqrt{1+\rho_{AB}\rho_{AB}^*}}\begin{bmatrix} 1 \\ \rho_{AB} \end{bmatrix}$$

where $|\mathbf{E}| = \sqrt{E_A E_A^* + E_B E_B^*}$ is the amplitude of the wave $\mathbf{E}(AB)$. If we choose $|\mathbf{E}| = 1$ and disregard the absolute phase ϕ_A, the above representation becomes

$$\mathbf{E}(AB) = \frac{1}{\sqrt{1+\rho_{AB}\rho_{AB}^*}}\begin{bmatrix} 1 \\ \rho_{AB} \end{bmatrix} \tag{2.28}$$

This representation of the polarization state using the polarization ratio ρ_{AB} is very useful. For example, if we want to represent a left-handed circular (LHC) polarization state and a right-handed circular (RHC) polarization state in a linear basis {H V} using the polarization ratio. For a left-handed circular (LHC) polarization, $|E_H| = |E_V|$, $\phi_{HV} = \phi_V - \phi_H = \pi/2$, and according to (2.26), the polarization ratio ρ_{HV} is j. Using (2.28) with $\rho_{HV} = j$, we obtain for the left-handed circular (LHC) polarization

$$\mathbf{E}(HV) = \frac{1}{\sqrt{2}}\begin{bmatrix} 1 \\ j \end{bmatrix} \tag{2.29}$$

Similarly, the polarization ratio ρ_{HV} of a right-handed circular (RHC) polarization state in a linear basis {H V} is $-j$ because the relative phase $\phi_{HV} = -\pi/2$, and its representation is

$$\mathbf{E}(HV) = \frac{1}{\sqrt{2}}\begin{bmatrix} 1 \\ -j \end{bmatrix} \tag{2.30}$$

The complex polarization ratio ρ is important in radar polarimetry. However, the value of the polarization ratio ρ defined in a certain polarization basis is different from that defined in the other polarization basis even if the physical polarization state is the same.

2.4.1 Complex Polarization Ratio in the Linear Basis {H V}

In the linear {H V} basis with unit vectors $\hat{h}$ and $\hat{v}$, a polarization state may be expressed as:

$$\mathbf{E}(HV) = E_H\,\hat{h} + E_V\,\hat{v} \tag{2.31}$$

The polarization ratio ρ_{HV}, according to (2.6), can be described as:

$$\rho_{HV} = \frac{E_V}{E_H} = \left|\frac{E_V}{E_H}\right| \exp\left(j\phi_{HV}\right) = \tan\alpha_{HV}\exp\left(j\phi_{HV}\right), \qquad \phi_{HV} = \phi_V - \phi_H \tag{2.32}$$

where the angle α_{HV} is defined in Fig. 2.1c, only in the {H V} basis and

$$\begin{aligned} |E_H| &= \sqrt{E_H^2 + E_V^2}\cos\alpha_{HV} \\ |E_V| &= \sqrt{E_H^2 + E_V^2}\sin\alpha_{HV} \end{aligned} \tag{2.33}$$

Also, for a single monochromatic, uniform TEM (transverse electromagnetic) traveling plane wave in the positive z direction, the real instantaneous electric field is written as

$$\boldsymbol{\varepsilon}(z,t) = \begin{bmatrix} \varepsilon_x(z,t) \\ \varepsilon_y(z,t) \\ \varepsilon_z(z,t) \end{bmatrix} = \begin{bmatrix} |E_x|\cos(\omega t - kz + \phi_x) \\ |E_y|\cos(\omega t - kz + \phi_y) \\ 0 \end{bmatrix} \tag{2.34}$$

In a cartesian coordinate system, the $+x$-axis is commonly chosen as the horizontal basis (H) and the $+y$-axis as the vertical basis (V) Substituting (2.33) into (2.34), we find

$$\begin{aligned} \boldsymbol{\varepsilon}(z,t) &= \begin{bmatrix} \sqrt{E_H^2 + E_V^2}\cos\alpha_{HV}\cos(\omega t - kz + \phi_H) \\ \sqrt{E_H^2 + E_V^2}\sin\alpha_{HV}\cos(\omega t - kz + \phi_V) \end{bmatrix} = \\ &= \sqrt{E_H^2 + E_V^2}\exp\left\{\begin{bmatrix} \cos\alpha_{HV} \\ \sin\alpha_{HV}\exp(j\phi) \end{bmatrix}\exp\{j(\omega t - kz + \phi_H)\}\right\} \end{aligned} \tag{2.35}$$

where $\phi = \phi_V - \phi_H$ is the relative phase. The expression in the square bracket is a spinor [32] which is independent of the time-space dependence of the traveling wave. The spinor parameters (α, ϕ) are easy to be located on the Poincaré sphere and can be used to represent the polarization state of a plane wave. In Fig. 2.4c, the polarization state, described by the point P_E on the Poincaré sphere, can be expressed in terms of these two angles, where $2\alpha_{HV}$ is the angle subtended by the great circle drawn from the point P_E on the equator measured from H toward V; and ϕ_{HV} is the angle between the great circle and the equator.

From equations, (2.7) and (2.8) for the {H V} basis we have

$$\begin{aligned} \sin 2\chi &= \sin 2\alpha_{HV} \sin \phi_{HV} \\ \tan 2\psi &= \tan(2\alpha_{HV}) \cos \phi_{HV} \end{aligned} \tag{2.36}$$

which describes the ellipticity angle χ and the tilt or orientation angle ψ in terms of the variables α_{HV} and ϕ_{HV}. Also, from (2.11) for the {H V} basis an inverse pair that describes the α_{HV} and ϕ_{HV} in terms of χ and ψ is given in (2.37)

$$\begin{aligned} \cos 2\alpha_{HV} &= \cos 2\psi \cos 2\chi \\ \tan \phi_{HV} &= \frac{\tan 2\chi}{\sin 2\psi} \end{aligned} \tag{2.37}$$

It is convenient to describe the polarization state by either of the two set of angles (α_{HV}, ϕ_{HV}) or (χ, ψ) which describe a point on the Poincaré sphere. The complex polarization ratio ρ_{HV} can be used to specify the polarization of an electromagnetic wave expressed in the {H V} basis. Some common polarization states expressed in terms of (χ, ψ), ρ, and the normalized Jones vector $\mathbf{E}$ are listed in Table 2.1 at the end of this section.

2.4.2 Complex Polarization Ratio in the Circular Basis {L R}

In the circular basis {L R}, we have two unit vectors $\hat{\mathrm{L}}$ (left-handed circular) and $\hat{\mathrm{R}}$ (right- handed circular). Any polarization of a plane wave can be expressed by

$$\mathbf{E}(LR) = E_L \hat{\mathrm{L}} + E_R \hat{\mathrm{R}} \tag{2.38}$$

A unit amplitude left-handed circular polarization has only the L component in the circular basis {L R}. It can be expressed by

$$\mathbf{E}(LR) = 1 * \hat{\mathrm{L}} + 0 * \hat{\mathrm{R}} = \begin{bmatrix} 1 \\ 0 \end{bmatrix} \tag{2.39}$$

The above representation of a unit (LHC) polarization in the circular basis {L R} is different from that in the linear basis {H V} of (2.29). Similarly, a unit amplitude right-handed circular polarization has only the R component in the circular basis {L R}

$$\mathbf{E}(LR) = 0 * \hat{\mathrm{L}} + 1 * \hat{\mathrm{R}} = \begin{bmatrix} 0 \\ 1 \end{bmatrix} \tag{2.40}$$

which is different from that in the linear {H V} basis.

The polarization ratio ρ_{LR}, according to (2.26) is

$$\rho_{LR} = \frac{E_R}{E_L} = \frac{|E_R|}{|E_L|}\exp\{j(\phi_R - \phi_L)\} = |\rho_{LR}|\exp\{j\phi_{LR}\} = \tan\alpha_{LR}\exp\{j\phi_{LR}\} \tag{2.41}$$

where $|\rho_{LR}|$ is the ratio of magnitudes of the two orthogonal components $|E_L|$ and $|E_R|$, and ϕ_{LR} the phase difference. The angles α_{LR} and ϕ_{LR} are also easy to be found on the Poincaré sphere (see Fig. 2.6) like the angles α_{HV} and ϕ_{HV}. Some common polarization states in terms of ρ_{LR}, are listed in Table 2.1.

2.4.3 Complex Polarization Ratio in the Linear Basis {45° 135°}

In the linear {45° 135°} basis with unit vectors $\hat{45}^\circ$ and $\hat{135}^\circ$, a polarization state may be expressed as

$$\mathbf{E}(45^\circ 135^\circ) = E_{45^\circ}\hat{45}^\circ + E_{135^\circ}\hat{135}^\circ \tag{2.42}$$

where E_{45° and E_{135° are the 45° component and the 135° component, respectively. The polarization ratio according to (2.26) is

$$\rho_{45^\circ 135^\circ} = \frac{E_{135^\circ}}{E_{45^\circ}} = \frac{|E_{135^\circ}|}{|E_{45^\circ}|}\exp\{j(\phi_{135^\circ} - \phi_{45^\circ})\} = |\rho_{45^\circ 135^\circ}|\exp\{j\phi_{45^\circ 135^\circ}\} = \tan\alpha_{45^\circ 135^\circ}\exp\{j\phi_{45^\circ 135^\circ}\} \tag{2.43}$$

where $|\rho_{45^\circ 135^\circ}|$ is the ratio of magnitudes of the two orthogonal components $|E_{135^\circ}|$ and $|E_{45^\circ}|$, and $\phi_{45^\circ 135^\circ}$ the phase difference. The angles $\alpha_{45^\circ 135^\circ}$ and $\phi_{45^\circ 135^\circ}$ are also easy to be found on the Poincaré sphere (see Fig. 2.6)

TABLE 2.1
POLARIZATION STATES IN TERMS OF (χ, ψ), POLARIZATION RATIO ρ AND NORMALIZED JONES VECTOR **E**

POLARIZATION	χ	ψ	{H V} basis		{45° 135°} basis		{L R} basis	
			ρ_{HV}	**E**	$\rho_{45^\circ 135^\circ}$	**E**	ρ_{LR}	**E**
Linear Horizontal	0	0	0	$\begin{bmatrix}1\\0\end{bmatrix}$	-1	$\frac{1}{\sqrt{2}}\begin{bmatrix}1\\-1\end{bmatrix}$	1	$\frac{1}{\sqrt{2}}\begin{bmatrix}1\\1\end{bmatrix}$
Linear Vertical	0	$\frac{\pi}{2}$	∞	$\begin{bmatrix}0\\1\end{bmatrix}$	1	$\frac{1}{\sqrt{2}}\begin{bmatrix}1\\1\end{bmatrix}$	-1	$\frac{1}{\sqrt{2}}\begin{bmatrix}-j\\j\end{bmatrix}$
45° Linear	0	$\frac{\pi}{4}$	1	$\frac{1}{\sqrt{2}}\begin{bmatrix}1\\1\end{bmatrix}$	0	$\begin{bmatrix}1\\0\end{bmatrix}$	j	$\frac{1}{2}\begin{bmatrix}1 & -j\\1 & j\end{bmatrix}$
135° Linear	0	$-\frac{\pi}{4}$	-1	$\frac{1}{\sqrt{2}}\begin{bmatrix}-1\\1\end{bmatrix}$	∞	$\begin{bmatrix}0\\1\end{bmatrix}$	$-j$	$\frac{1}{2}\begin{bmatrix}-1 & -j\\-1 & j\end{bmatrix}$
Left-handed Circular	$\frac{\pi}{4}$		j	$\frac{1}{\sqrt{2}}\begin{bmatrix}1\\j\end{bmatrix}$	j	$\frac{1}{2}\begin{bmatrix}1 & j\\-1 & j\end{bmatrix}$	0	$\begin{bmatrix}1\\0\end{bmatrix}$
Right-handed Circular	$-\frac{\pi}{4}$		$-j$	$\frac{1}{\sqrt{2}}\begin{bmatrix}1\\-j\end{bmatrix}$	$-j$	$\frac{1}{2}\begin{bmatrix}1 & -j\\-1 & -j\end{bmatrix}$	∞	$\begin{bmatrix}0\\1\end{bmatrix}$

2.5 The Stokes Parameters

So far, we have seen completely polarized waves for which $|E_A|, |E_B|$, and ϕ_{AB} are constants or at least slowly varying functions of time. If we need to deal with partial polarization, it is convenient to use the Stokes parameters q_0, q_1, q_2 *and* q_3 introduced by Stokes in 1852 [107] for describing partially polarized waves by observable power terms and not by amplitudes (and phases).

2.5.1 The Stokes vector for the completely polarized wave

For a monochromatic wave, in the linear {H V} basis, the four Stokes parameters are

$$\begin{aligned} q_0 &= |E_H|^2 + |E_V|^2 \\ q_1 &= |E_H|^2 - |E_V|^2 \\ q_2 &= 2|E_H||E_V|\cos\phi_{HV} \\ q_3 &= 2|E_H||E_V|\sin\phi_{HV} \end{aligned} \tag{2.44}$$

For a completely polarized wave, there are only three independent parameters, which are related as follows

$$q_0^2 = q_1^2 + q_2^2 + q_3^2 \tag{2.45}$$

The Stokes parameters are sufficient to characterize the magnitude and the relative phase, and hence the polarization of a wave. The Stokes parameter q_0 is always equal to the total power (density) of the wave; q_1 is equal to the power in the linear horizontal or vertical polarized components; q_2 is equal to the power in the linearly polarized components at tilt angles $\psi = 45^\circ$ or 135°; and q_3 is equal to the power in the left-handed and right-handed circular polarized components. If any of the parameters q_0, q_1, q_2 or q_3 has a non-zero value, it indicates the presence of a polarized component in the plane wave. The Stokes parameters are also related to the geometric parameters A, χ, and ψ of the polarization ellipse

$$\mathbf{q} = \begin{bmatrix} q_0 \\ q_1 \\ q_2 \\ q_3 \end{bmatrix} = \begin{bmatrix} |E_H|^2 + |E_V|^2 \\ |E_H|^2 - |E_V|^2 \\ 2|E_H||E_V|\cos\phi_{HV} \\ 2|E_H||E_V|\sin\phi_{HV} \end{bmatrix} = \begin{bmatrix} A^2 \\ A^2\cos 2\psi\cos 2\chi \\ A^2\sin 2\psi\cos 2\chi \\ A^2\sin 2\chi \end{bmatrix} \tag{2.46}$$

which for the normalized case $q_0^2 = e^2 = e_H^2 + e_V^2 = 1$ and

$$\mathbf{q} = \begin{bmatrix} q_0 \\ q_1 \\ q_2 \\ q_3 \end{bmatrix} = \frac{1}{\sqrt{|E_H|^2 + |E_V|^2}} \begin{bmatrix} |E_H|^2 + |E_V|^2 \\ |E_H|^2 - |E_V|^2 \\ 2\operatorname{Re}\{E_H^* E_V\} \\ 2\operatorname{Im}\{E_H^* E_V\} \end{bmatrix} = \begin{bmatrix} e_H^2 + e_V^2 \\ e_H^2 - e_V^2 \\ 2e_H e_V\cos\phi \\ 2e_H e_V\sin\phi \end{bmatrix} = \begin{bmatrix} e^2 \\ e^2\cos 2\psi\cos 2\chi \\ e^2\sin 2\psi\cos 2\chi \\ e^2\sin 2\chi \end{bmatrix} \tag{2.47}$$

2.5.2 The Stokes vector for the partially polarized wave

The Stokes parameter presentation [34] possesses two main advantages in that all of the four parameters are measured as intensities, a crucial fact in optical polarimetry, and the ability to present partially polarized waves in terms of the 2x2 complex hermitian positive semi-definite wave coherency matrix $[J]$ also called the Wolf's coherence matrix [34], defined as:

$$[J]=\langle \mathbf{E}\mathbf{E}^{\dagger}\rangle=\begin{bmatrix}\langle E_H E_H^*\rangle & \langle E_H E_V^*\rangle\\ \langle E_V E_H^*\rangle & \langle E_V E_V^*\rangle\end{bmatrix}=\begin{bmatrix}J_{HH} & J_{HV}\\ J_{VH} & J_{VV}\end{bmatrix}=\begin{bmatrix}q_0+q_1 & q_2+jq_3\\ q_2-jq_3 & q_0-q_1\end{bmatrix} \tag{2.48}$$

where $\langle ...\rangle=\lim_{T\to\infty}\left[\frac{1}{2T}\int_{-T}^{T}\langle ...\rangle dt\right]$ indicating temporal or ensemble averaging assuming stationarity of the wave. We can associate the Stokes vector $\mathbf{q}$ with the coherency matrix $[J]$

$$\begin{aligned}
q_0&=|E_H|^2+|E_V|^2=\langle E_H E_H^*\rangle+\langle E_V E_V^*\rangle=J_{HH}+J_{VV}\\
q_1&=|E_H|^2-|E_V|^2=\langle E_H E_H^*\rangle-\langle E_V E_V^*\rangle=J_{HH}-J_{VV}\\
q_2&=2|E_H||E_V|\cos\phi_{HV}=\langle E_H E_V^*\rangle+\langle E_V E_H^*\rangle=J_{HV}+J_{VH}\\
q_3&=2|E_H||E_V|\sin\phi_{HV}=j\langle E_H E_V^*\rangle-j\langle E_V E_H^*\rangle=jJ_{HV}-jJ_{VH}
\end{aligned} \tag{2.49}$$

and since $[J]$ is positive semidefinite matrix

$$Det\{[J]\}\geq 0 \;\; or \;\; q_0^2\geq q_1^2+q_2^2+q_3^2 \tag{2.50}$$

the diagonal elements presenting the intensities, the off-diagonal elements the complex cross-correlation between E_H and E_V, and the $Trace\{[J]\}$, representing the total energy of the wave. For $J_{HV}=0$ no correlation between E_H and E_V exists, $[J]$ is diagonal with $J_{HH}=J_{VV}$, (i.e. the wave is unpolarized or completely depolarized, and possesses *one degree of freedom only: amplitude*). Whereas, for $Det\{[J]\}=0$ we find that $J_{VH}J_{HV}=J_{HH}J_{VV}$, and the correlation between E_H and E_V is maximum, and the wave is completely polarized in which case the wave *possesses three degrees of freedom: amplitude, orientation, and ellipticity of the polarization ellipse*. Between these two extreme cases lies the general case of partial polarization, where $Det\{[J]\}>0$ is indicating a certain degree of statistical dependence between E_H and E_V which can be expressed in terms of the '*degree of coherency*' μ and the 'degree *of polarization*' D_p as

$$\mu_{HV}=|\mu_{HV}|\exp(j\beta_{HV})=\frac{J_{HV}}{\sqrt{J_{HH}J_{VV}}} \tag{2.51}$$

$$D_p=\left(1-\frac{4Det\{[J]\}}{(Trace\{[J]\})^2}\right)^{1/2}=\frac{\left(q_1^2+q_2^2+q_3^2\right)^{1/2}}{q_0} \tag{2.52}$$

where $\mu=D_p=0$ for totally depolarized and $\mu=D_p=1$ for fully polarized waves, respectively. However, under a change of polarization basis the elements of the wave coherency matrix $[J]$ depend on the choice of the polarization basis, where according to [52, 53], $[J]$ transforms through a unitary similarity transformation as

$$\langle [J_{ij}]\rangle=[U_2]\langle [J_{mn}]\rangle[U_2]^{\dagger} \tag{2.53}$$

The fact that the trace and the determinant of a hermitian matrix are invariant under unitary similarity transformations means that both, the degree of polarization as well as the total wave intensity are not affected by polarimetric basis transformations. Also, note that the degree of coherence μ_{mn} does depend on the polarization basis. Table 2.2 gives the Jones vector $\mathbf{E}$, Coherency Matrix $[J]$, and Stokes Vector $\mathbf{q}$ for special cases of purely monochromatic wave fields in specific states of polarization.

TABLE 2.2
JONES VECTOR $\mathbf{E}$, COHERENCY MATRIX $[J]$, AND STOKES VECTOR $\mathbf{q}$ FOR SOME STATES OF POLARIZATION

POLARIZATION	{H V} BASIS		
	$\mathbf{E}$	$[J]$	$\mathbf{q}$
Linear Horizontal	$\begin{bmatrix}1\\0\end{bmatrix}$	$\begin{bmatrix}1&0\\0&0\end{bmatrix}$	$\begin{bmatrix}1\\1\\0\\0\end{bmatrix}$
Linear Vertical	$\begin{bmatrix}0\\1\end{bmatrix}$	$\begin{bmatrix}0&0\\0&1\end{bmatrix}$	$\begin{bmatrix}1\\-1\\0\\0\end{bmatrix}$
45° Linear	$\frac{1}{\sqrt{2}}\begin{bmatrix}1\\1\end{bmatrix}$	$\frac{1}{2}\begin{bmatrix}1&1\\1&1\end{bmatrix}$	$\begin{bmatrix}1\\0\\1\\0\end{bmatrix}$
135° Linear	$\frac{1}{\sqrt{2}}\begin{bmatrix}-1\\1\end{bmatrix}$	$\frac{1}{2}\begin{bmatrix}1&-1\\-1&1\end{bmatrix}$	$\begin{bmatrix}1\\0\\-1\\0\end{bmatrix}$
Left-handed Circular	$\frac{1}{\sqrt{2}}\begin{bmatrix}1\\j\end{bmatrix}$	$\frac{1}{2}\begin{bmatrix}1&-j\\j&1\end{bmatrix}$	$\begin{bmatrix}1\\0\\0\\1\end{bmatrix}$
Right-handed Circular	$\frac{1}{\sqrt{2}}\begin{bmatrix}1\\-j\end{bmatrix}$	$\frac{1}{2}\begin{bmatrix}1&j\\-j&1\end{bmatrix}$	$\begin{bmatrix}1\\0\\0\\-1\end{bmatrix}$

2.6 The Poincaré Polarization Sphere

The Poincaré sphere, shown in Fig. 2.4 for the representation of wave polarization using the Stokes vector and the Deschamps parameters (α, ϕ) is a useful graphical aid for the visualization of polarization effects. There is one-to-one correspondence between all possible polarization states and points on the Poincaré sphere, with the linear polarizations mapped onto the equatorial plane (x = 0) with the *'zenith'* presenting left-handed circular and the *'nadir'* right-handed circular polarization states according to the IEEE standard notation $\exp(+j\omega t)$ [102], and any set of orthogonally fully polarized polarization states being mapped into antipodal points on the Poincaré sphere [108].

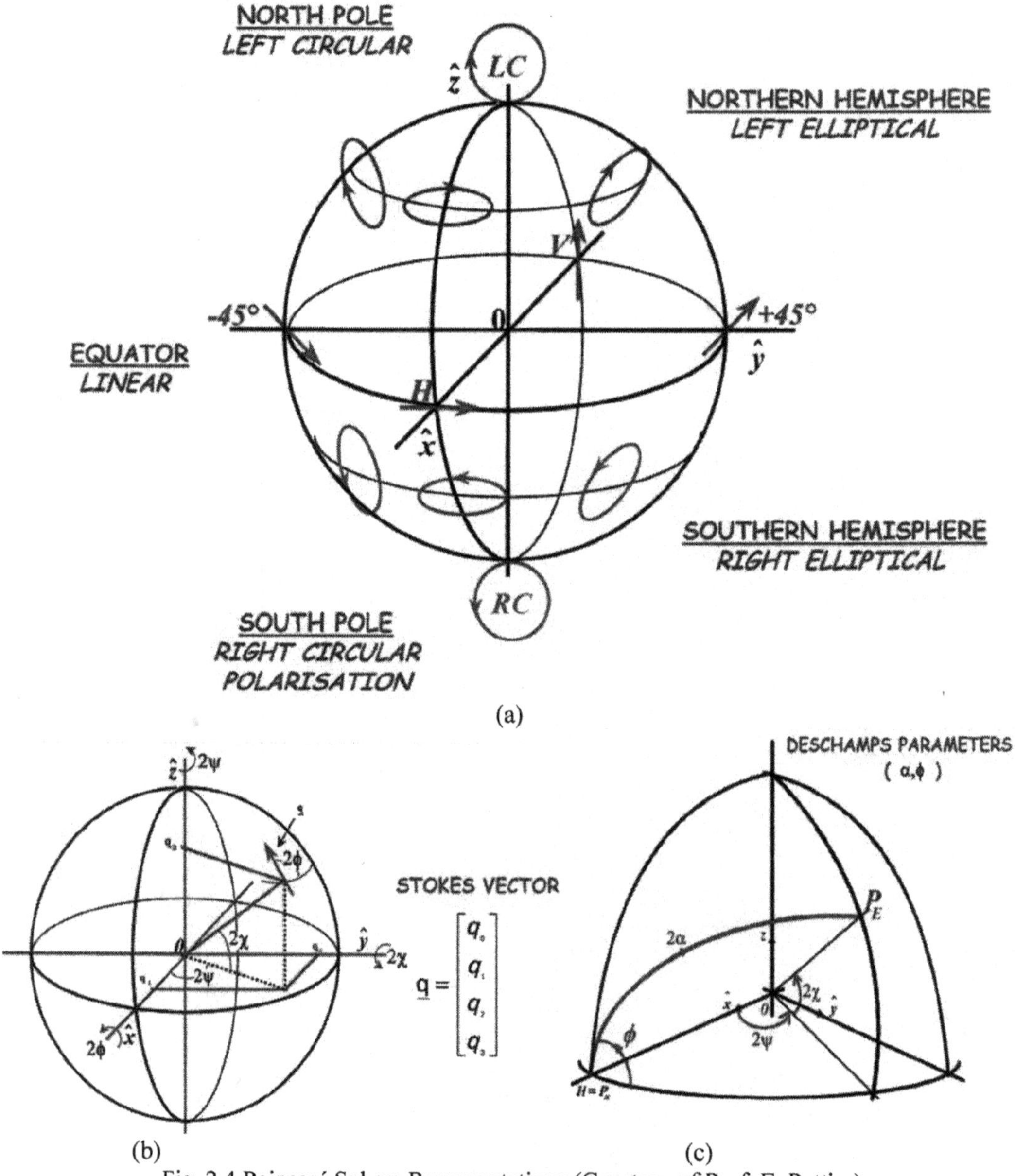

Fig. 2.4 Poincaré Sphere Representations (Courtesy of Prof. E. Pottier)

2.6.1 The polarization state on the Poincaré sphere for the {H V} basis

In the Poincaré sphere representation, the polarization state is described by a point P on the sphere, where the three Cartesian coordinate components are q_1, q_2, and q_3 according to (2.46). So, for any state of a completely polarized wave, there corresponds one point $P(q_1, q_2, q_3)$ on the sphere of radius q_0, and vice versa. In Fig. 2.5, we can see that the longitude and latitude of the point P are related to the geometric parameter of the polarization ellipse and they are 2ψ and 2χ respectively.

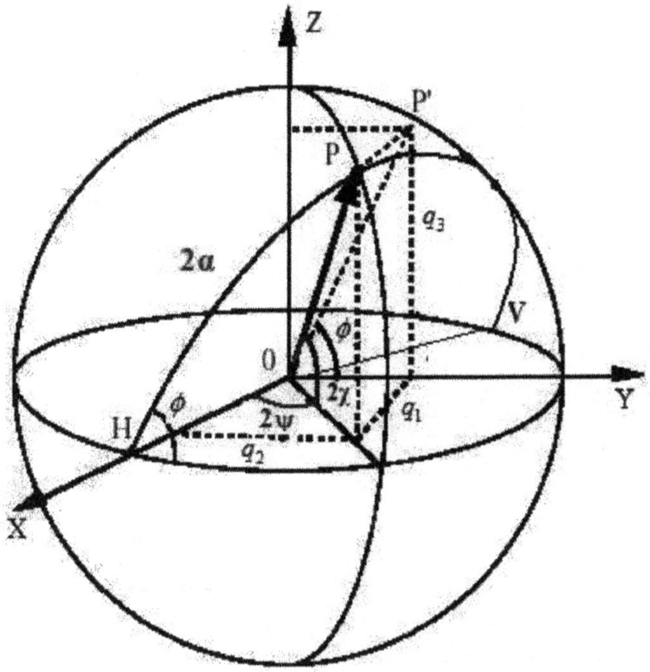

Fig. 2.5 The Poincaré sphere and the parameters α_{HV} and ϕ_{HV}

In addition, the point P on the Poincaré sphere can also be represented by the angles α_{HV} and ϕ_{HV}. From (2.37) and (2.46) we find that

$$\frac{q_1}{q_0} = \cos 2\psi \cos 2\chi = \cos 2\alpha_{HV} \tag{2.54}$$

where $\cos 2\alpha_{HV}$ is the direction cosine of the Stokes vector $\mathbf{q}$ with respect to the X-axis, i.e., the angle $2\alpha_{HV}$ is the angle between $\mathbf{q}$ and the X-axis. The angle ϕ_{HV} is the angle between the equator and the great circle with basis diameter HV through the point P, and it is equal to the angle between the XOY plane and the XOP plane. Drawing a projecting line from point P to the YOZ plane, the intersecting point P' is on the XOP plane, so $\phi_{HV} = \angle \mathrm{YOP}'$ ($\phi_{HV} = \phi$ in Fig. 2.5). On the YOZ plane we find that

$$\tan \phi_{HV} = \tan \angle \mathrm{YOP}' = \frac{q_3}{q_2} = \frac{\tan 2\chi}{\sin 2\psi} \tag{2.55}$$

which satisfies equations (2.46) and (2.37).

2.6.2 The polarization ratio on the Poincaré sphere for different polarization bases

Also, it can be shown that a polarization state can be represented in different polarization bases. Any polarization basis consists of two unit vectors which are located at two corresponding antipodal points on the Poincaré sphere. Fig 2.6 shows how the polarization state P on the Poincaré sphere can be represented in three polarization bases, {H V}, {45° 135°}, and {L R}. The complex polarization ratios are given by

$$\rho_{HV} = |\rho_{HV}| \exp(j\phi_{HV}) = \begin{cases} \tan\alpha_{HV} \exp(j\phi_{HV}) & 0 < \alpha_{HV} < \frac{\pi}{2} \\ -\tan\alpha_{HV} \exp(j\phi_{HV}) & \frac{\pi}{2} < \alpha_{HV} < \pi \end{cases} \tag{2.56}$$

$$\rho_{45°135°} = |\rho_{45°135°}| \exp(j\phi_{45°135°}) = \begin{cases} \tan\alpha_{45°135°} \exp(j\phi_{45°135°}) & 0 < \alpha_{45°135°} < \frac{\pi}{2} \\ -\tan\alpha_{45°135°} \exp(j\phi_{45°135°}) & \frac{\pi}{2} < \alpha_{45°135°} < \pi \end{cases} \tag{2.57}$$

$$\rho_{LR} = |\rho_{LR}| \exp(j\phi_{LR}) = \begin{cases} \tan\alpha_{LR} \exp(j\phi_{LR}) & 0 < \alpha_{LR} < \frac{\pi}{2} \\ -\tan\alpha_{LR} \exp(j\phi_{LR}) & \frac{\pi}{2} < \alpha_{LR} < \pi \end{cases} \tag{2.58}$$

where $\tan\alpha_{HV}$, $\tan\alpha_{45°135°}$, and $\tan\alpha_{LR}$ are the ratios of the magnitudes of the corresponding orthogonal components, and ϕ_{HV}, $\phi_{45°135°}$, and ϕ_{LR} are the phase differences between the corresponding orthogonal components

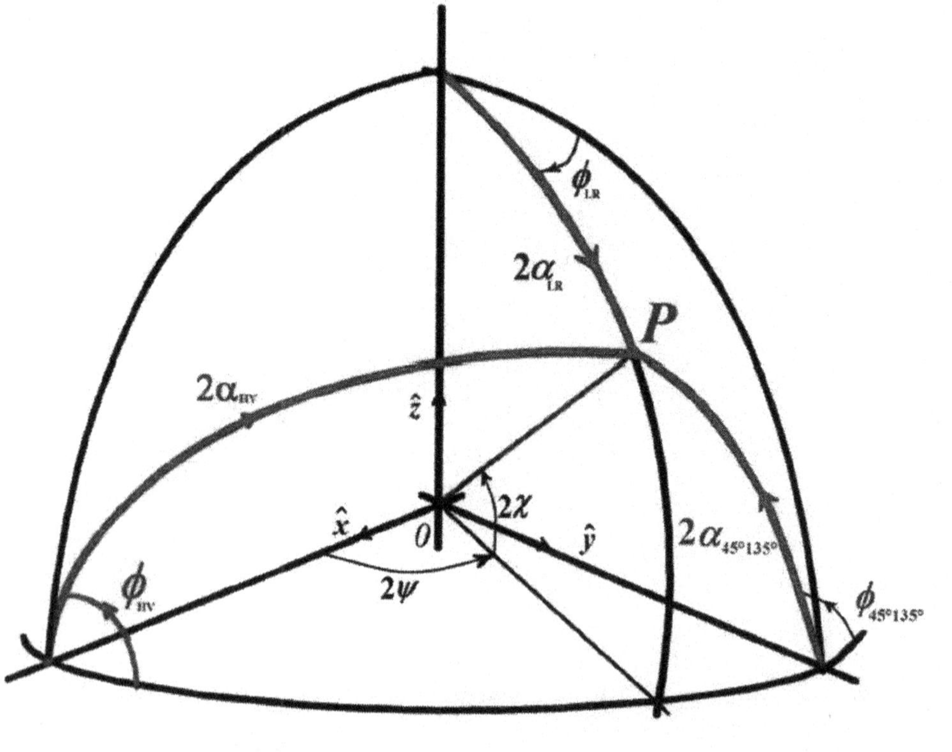

Fig 2.6 The Polarization State P in Different Polarization Bases

2.6.3 The relationship between the Stokes vector and the polarization ratio for different polarization bases

First, consider the polarization ratio ρ_{HV} defined in the {H V} basis. Because $\cos 2\alpha_{HV}$ is the direction cosine of the Stokes vector **q** with respect to the X-axis, we find

$$\frac{q_1}{q_0} = \cos 2\alpha_{HV} = \frac{1-\tan^2\alpha_{HV}}{1+\tan^2\alpha_{HV}} = \frac{1-|\rho_{HV}|^2}{1+|\rho_{HV}|^2} \tag{2.59}$$

the straight forward solution for $|\rho_{HV}|$ is

$$|\rho_{HV}| = \sqrt{\frac{q_0-q_1}{q_0+q_1}} \tag{2.60}$$

from (2.54), we find

$$\phi_{HV} = \angle \mathrm{YOP'} = \tan^{-1}\left(\frac{q_3}{q_2}\right) \tag{2.61}$$

Combining above two equations yields

$$\rho_{HV} = |\rho_{HV}|\exp(j\phi_{HV}) = \sqrt{\frac{q_0-q_1}{q_0+q_1}}\exp\left\{j\tan^{-1}\left(\frac{q_3}{q_2}\right)\right\} \tag{2.62}$$

For a completely polarized wave, we may obtain the Stokes vector in terms of the polarization ratio ρ_{HV} by applying

$$\begin{aligned}
q_0 &= \sqrt{q_1^2+q_2^2+q_3^2} = 1\\
q_1 &= \frac{1-|\rho_{HV}|^2}{1+|\rho_{HV}|^2} = \cos 2\alpha_{HV}\\
q_2 &= \frac{2|\rho_{HV}|\cos\phi_{HV}}{1+|\rho_{HV}|^2} = \frac{2\tan\alpha_{HV}\cos\phi_{HV}}{1+|\tan\alpha_{HV}|^2} = \sin(2\alpha_{HV})\cos\phi_{HV}\\
q_3 &= \frac{2|\rho_{HV}|\sin\phi_{HV}}{1+|\rho_{HV}|^2} = \sin(2\alpha_{HV})\sin\phi_{HV}
\end{aligned} \tag{2.63}$$

The sign of the three components of the Stokes vector is summarized in Table 2.3.

Secondly, consider the polarization ratio $\rho_{45°135°}$ defined in the {45° 135°} basis. The $\cos 2\alpha_{45°135°}$ is the direction cosine of the Stokes vector **q** with respect to the Y-axis. So similarly, with

$$
\begin{aligned}
&q_0 = 1 \\
&\left|\rho_{45^\circ 135^\circ}\right| = \sqrt{\frac{q_0 - q_2}{q_0 + q_2}} \\
&\phi_{45^\circ 135^\circ} = \tan^{-1}\left(-\frac{q_3}{q_1}\right)
\end{aligned}
\tag{2.64}
$$

TABLE 2.3

THE SIGN OF THE q_1, q_2, AND q_3 PARAMETERS IN THE {H V} BASIS

ϕ_{HV}	α_{HV}	q_1	q_2	q_3
$0 < \phi_{HV} < \frac{\pi}{2}$	$0 < 2\alpha_{HV} < \frac{\pi}{2}$	+	+	+
	$\frac{\pi}{2} < 2\alpha_{HV} < \pi$	−	+	+
	$0 < 2\alpha_{HV} < \frac{3\pi}{2}$	−	−	−
	$\frac{3\pi}{2} < 2\alpha_{HV} < 2\pi$	+	−	−
$-\frac{\pi}{2} < \phi_{HV} < 0$	$0 < 2\alpha_{HV} < \frac{\pi}{2}$	+	+	−
	$\frac{\pi}{2} < 2\alpha_{HV} < \pi$	−	+	−
	$0 < 2\alpha_{HV} < \frac{3\pi}{2}$	−	−	+
	$\frac{3\pi}{2} < 2\alpha_{HV} < 2\pi$	+	−	+

Then the polarization ratio $\rho_{45^\circ 135^\circ}$ can be determined by the Stokes vector $\mathbf{q}$

$$
\rho_{45^\circ 135^\circ} = \sqrt{\frac{q_0 - q_2}{q_0 + q_2}} \exp\left\{ j \tan^{-1}\left(-\frac{q_3}{q_1}\right)\right\}
\tag{2.65}
$$

Also, the Stokes vector $\mathbf{q}$ can be determined by the polarization ratio $\rho_{45^\circ 135^\circ}$ as follows:

$$
\begin{aligned}
&q_0 = 1 \\
&q_1 = \frac{2\left|\rho_{45^\circ 135^\circ}\right| \cos\phi_{45^\circ 135^\circ}}{1 + \left|\rho_{45^\circ 135^\circ}\right|^2} = -\sin 2\alpha_{45^\circ 135^\circ} \cos\phi_{45^\circ 135^\circ} \\
&q_2 = \frac{1 - \left|\rho_{45^\circ 135^\circ}\right|^2}{1 + \left|\rho_{45^\circ 135^\circ}\right|^2} = \cos 2\alpha_{45^\circ 135^\circ} \\
&q_3 = \frac{2\left|\rho_{45^\circ 135^\circ}\right| \sin\phi_{45^\circ 135^\circ}}{1 + \left|\rho_{45^\circ 135^\circ}\right|^2} = \sin 2\alpha_{45^\circ 135^\circ} \sin\phi_{45^\circ 135^\circ}
\end{aligned}
\tag{2.66}
$$

Finally, consider the polarization ratio ρ_{LR} defined in the {L R} basis. Similarly, because the $\cos 2\alpha_{LR}$ is the direction cosine of the Stokes vector $\mathbf{q}$ with respect to the Z-axis, the polarization ratio ρ_{LR} can be determined by the Stokes vector $\mathbf{q}$ as:

$$\rho_{LR} = \sqrt{\frac{q_0 - q_3}{q_0 + q_3}} \exp\left\{ j \tan^{-1}\left(\frac{q_2}{q_1} \right) \right\} \tag{2.67}$$

Inversely,

$$\begin{aligned} q_0 &= 1 \\ q_1 &= \frac{2|\rho_{LR}|\cos\phi_{LR}}{1+|\rho_{LR}|^2} = \sin 2\alpha_{LR}\cos\phi_{LR} \\ q_2 &= \frac{2|\rho_{LR}|\sin\phi_{LR}}{1+|\rho_{LR}|^2} = \sin 2\alpha_{LR}\sin\phi_{LR} \\ q_3 &= \frac{1-|\rho_{LR}|^2}{1+|\rho_{LR}|^2} = \cos 2\alpha_{LR} \end{aligned} \tag{2.68}$$

TABLE 2.4
ALTERNATE EXPRESSIONS FOR NORMALIZED STOKES VECTOR PRESENTATIONS ON THE POLARIZATION SPHERE

	χ, ψ	α_{HV}, ϕ_{HV}	$\alpha_{45^\circ 135^\circ}, \phi_{45^\circ 135^\circ}$	α_{LR}, ϕ_{LR}
q_0	1	1	1	1
q_1	$\cos 2\chi \cos 2\psi$	$\cos 2\alpha_{HV}$	$-\sin 2\alpha_{45^\circ 135^\circ} \cos\phi_{45^\circ 135^\circ}$	$\sin 2\alpha_{LR}\cos\phi_{LR}$
q_2	$\cos 2\chi \sin 2\psi$	$\sin 2\alpha_{HV}\cos\phi_{HV}$	$\cos 2\alpha_{45^\circ 135^\circ}$	$\sin 2\alpha_{LR}\sin\phi_{LR}$
q_3	$\sin 2\chi$	$\sin 2\alpha_{HV}\sin\phi_{HV}$	$\sin 2\alpha_{45^\circ 135^\circ} \sin\phi_{45^\circ 135^\circ}$	$\cos 2\alpha_{LR}$

2.6.4 The Poincaré polarization sphere and complex polarization ratio plane

Using the Riemann transformation, Poincaré introduced the polarization sphere representation of Fig. 2.5 which gives a relationship between the polarization ratio ρ and its corresponding spherical coordinates on the Poincaré sphere. First we need to introduce an auxiliary complex parameter $\mathrm{u}(\rho)$, which is defined by the Riemann transformation [14] of the surface of the sphere onto the polar grid as follows,

$$\mathrm{u}(\rho) = \frac{1 - j\rho}{1 + j\rho} \tag{2.69}$$

in the {H V} basis, $\rho_{HV} = \tan\alpha_{HV} \exp\{j\phi_{HV}\} = \tan\alpha_{HV}(\cos\phi_{HV} + j\sin\phi_{HV})$, then

$$\mathrm{u} = \frac{(1 + \tan\alpha_{HV}\sin\phi_{HV}) - j\tan\alpha_{HV}\cos\phi_{HV}}{(1 - \tan\alpha_{HV}\sin\phi_{HV}) + j\tan\alpha_{HV}\cos\phi_{HV}}$$

$$|\mathrm{u}|^2 = \frac{1 + 2\tan\alpha_{HV}\sin\phi_{HV}) + \tan^2\alpha_{HV}}{1 - 2\tan\alpha_{HV}\sin\phi_{HV}) + \tan^2\alpha_{HV}}$$

$$\frac{|u|^2-1}{|u|^2+1}=\frac{2\tan\alpha_{HV}}{1+\tan^2\alpha_{HV}}\sin\phi_{HV}=\sin 2\alpha_{HV}\sin\phi_{HV}$$

according to (2.36) and Fig. 2.4b, the polar angle $\Theta=\pi/2-2\chi$ can be obtained from

$$\frac{|u|^2-1}{|u|^2+1}=\sin 2\chi=\sin(\pi/2-\Theta)=\cos\Theta$$

so that

$$\Theta=\cos^{-1}\left(\frac{|u|^2-1}{|u|^2+1}\right) \tag{2.70}$$

also, according to (2.36) and Fig. 2.4b, the spherical azimuthal angle $\Phi=2\psi$ can be obtained from $-\frac{\text{Im}\{u\}}{\text{Re}\{u\}}=\frac{2\tan\alpha_{HV}\cos\phi_{HV}}{1-\tan^2\alpha_{HV}}=\tan 2\psi=\tan\Phi$, so that the spherical azimuthal angle Φ becomes

$$\Phi=\tan^{-1}\left(\frac{\text{Im}\{u\}}{\text{Re}\{u\}}\right) \tag{2.71}$$

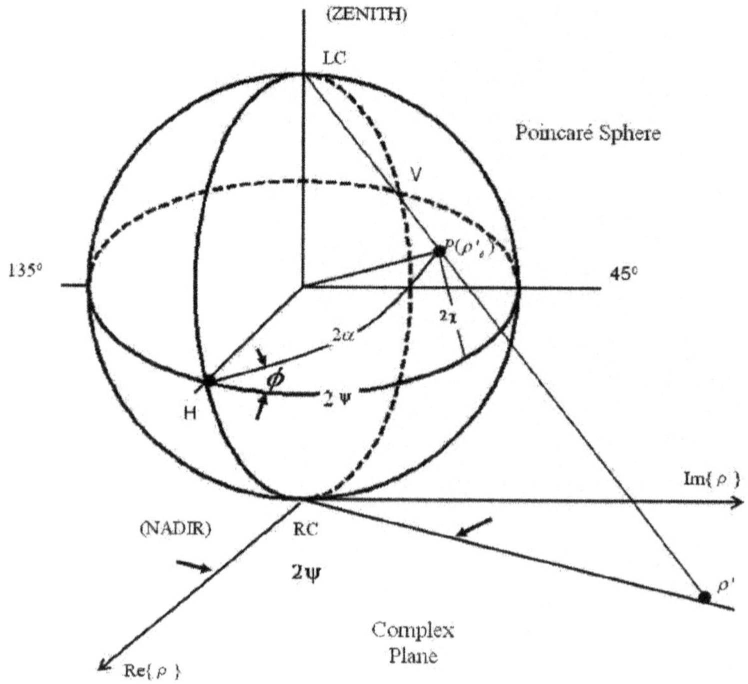

Fig. 2.7 Poincaré Sphere and the Complex Plane

2.7 Wave Decomposition Theorems

The diagonalization of $\left[J_{ij}\right]$, under the unitary similarity transformation is equivalent to finding an orthonormal polarization basis in which the coherency matrix is diagonal or

$$\begin{bmatrix} J_{mm} & J_{mn} \\ J_{nm} & J_{nn} \end{bmatrix} = \begin{bmatrix} e_{11} & e_{21} \\ e_{12} & e_{22} \end{bmatrix} \begin{bmatrix} \lambda_1 & 0 \\ 0 & \lambda_2 \end{bmatrix} \begin{bmatrix} e_{11}^* & e_{12}^* \\ e_{21}^* & e_{22}^* \end{bmatrix} \tag{2.72}$$

where λ_1 and λ_2 are the real non-negative eigenvalues of $[J]$ with $\lambda_1 \geq \lambda_2 \geq 0$, and $\hat{e}_1 = \begin{bmatrix} e_{11} & e_{12} \end{bmatrix}^T$ and $\hat{e}_2 = \begin{bmatrix} e_{21} & e_{22} \end{bmatrix}^T$ are the complex orthogonal eigenvectors which define $[U_2]$ and a polarization basis $\{\hat{\mathbf{e}}_1, \hat{\mathbf{e}}_2\}$ in which $[J]$ is diagonal. $[J]$ is Hermitian and hence normal and every normal matrix can be unitarily diagonalized . Being positive semidefinite the eigenvalues are nonnegative.

2.8 The Wave Dichotomy of Partially Polarized Waves

The solution of (2.72) provides two equivalent interpretations of partially polarized waves [28]: i) separation into fully polarized $[J_1]$, and into completely depolarized $[J_2]$ components

$$[J] = (\lambda_1 - \lambda_2)[J_1] + \lambda_2[I_2] \tag{2.72}$$

where $[I_2]$ is the 2x2 identity matrix ; ii) non-coherency of two orthogonal completely polarized wave states represented by the eigenvectors and weighed by their corresponding eigenvalues as

$$[J] = (\lambda_1)[J_1] + \lambda_2[J_2] = \lambda_1(\hat{\mathbf{e}}_1 \cdot \hat{\mathbf{e}}_1^\dagger) + \lambda_2(\hat{\mathbf{e}}_2 \cdot \hat{\mathbf{e}}_2^\dagger) \tag{2.74}$$

where $Det\{[J_1]\} = Det\{[J_2]\} = 0$; and if $\lambda_1 = \lambda_2$ the wave is totally depolarized (degenerate case) whereas for $\lambda_2 = 0$, the wave is completely polarized. Both models are unique in the sense that no other decomposition in form of a separation of two completely polarized waves or of a completely polarized with noise is possible for a given coherency matrix, which may be reformulated in terms of the *'degree of polarization'* D_p as

$$D_p = \frac{\lambda_1 - \lambda_2}{\lambda_1 + \lambda_2}, \quad 0 \ (\lambda_1 = \lambda_2) \ and \ 1 \ (\lambda_2 = 0) \tag{2.75}$$

for a partially unpolarized and completely polarized wave. The fact that the eigenvalues λ_1 and λ_2 are invariant under polarization basis transformation makes D_p an important basis-independent parameter.

2.9 Polarimetric Wave Entropy

Alternately to the degrees of wave coherency μ and polarization D_p, the polarimetric wave entropy H_ω [28] provides another measure of the correlated wave structure of the coherency matrix $[J]$, where by using the logarithmic sum of eigenvalues

$$H_\omega = \sum_{i=1}^{2} \{-P_i \log_2 P_i\} \quad with \ P_i = \frac{\lambda_i}{\lambda_1 + \lambda_2} \tag{2.76}$$

so that $P_1 + P_2 = 1$ and the normalized wave entropy ranges from $0 \leq H_\omega \leq 1$ where for a completely polarized wave with $\lambda_2 = 0$ and $H_\omega = 0$, while a completely randomly polarized wave with $\lambda_1 = \lambda_2$ possesses maximum entropy $H_\omega = 1$.

2.10 Alternate Formulations of the Polarization Properties of Electromagnetic Vector Waves

There exist several alternate formulations of the polarization properties of electromagnetic vector waves including; (i) the '*Four-vector Hamiltonian*' formulation frequently utilized by Zhivotovsky [109] and by Czyz [110], which may be useful in a more concise description of partially polarized waves ; (ii) the '*Spinorial formulation*' used by Bebbington [32], and in general gravitation theory [111] ; and (iii) a pseudo-spinorial formulation by Czyz [110] is in development which are most essential tools for describing the general bi-static (non-symmetric) scattering matrix cases for both the coherent (*3-D Poincaré sphere and the 3-D polarization spheroid*) and the partially polarized (*4-D Zhivotovsky sphere and 4-D spheroid)* cases [109]. Because of the exorbitant excessive additional mathematical tools required, and not commonly accessible to engineers and applied scientists, these formulations are not presented here but deserve our fullest attention in future analyses.

3. The Electromagnetic Vector Scattering Operator and the Polarimetric Scattering Matrices

The electromagnetic vector wave interrogation with material media is described by the Scattering Operator $[S(\mathbf{k}_s/\mathbf{k}_i)]$ with $\mathbf{k}_s,\mathbf{k}_i$ representing the wave vectors of the scattered and incident, $\mathbf{E}^s(\mathbf{r}),\mathbf{E}^i(\mathbf{r})$ respectively, where

$$\mathbf{E}^s(\mathbf{r}) = E_0^s \exp(-j\mathbf{k}_s\cdot\mathbf{r}) = \mathbf{e}_s E_0^s \exp(-j\mathbf{k}_s\cdot\mathbf{r}) \tag{3.1}$$

is related to

$$\mathbf{E}^i(\mathbf{r}) = E_0^i \exp(-j\mathbf{k}_i\cdot\mathbf{r}) = \mathbf{e}_i E_0^i \exp(-j\mathbf{k}_i\cdot\mathbf{r}) \tag{3.2}$$

$$\mathbf{E}^s(\mathbf{r}) = \frac{\exp(-j\mathbf{k}_s\cdot\mathbf{r})}{r}[S(\mathbf{k}_s/\mathbf{k}_i)]\mathbf{E}^i(\mathbf{r}) \tag{3.3}$$

The scattering operator $[S(\mathbf{k}_s/\mathbf{k}_i)]$ is obtained from rigorous application of vector scattering and diffraction theory, to the specific scattering scenario under investigation which is not further discussed here, but we refer to [97] for a thought-provoking formulation of these still open problems.

3.1 The Scattering Scenario and the Scattering Coordinate Framework

The scattering operator $[S(\mathbf{k}_s/\mathbf{k}_i)]$ appears as the output of the scattering process for an arbitrary input $\mathbf{E}_0^i$, which must carefully be defined in terms of the scattering scenario; and, its proper unique formulation is of intrinsic importance to both optical and radar polarimetry. Whereas in optical remote sensing mainly the '*forward scattering through translucent media*' is considered, in radar remote sensing the '*back scattering from distant, opaque open and closed surfaces*' is of interest, where in radar target backscattering we usually deal with closed surfaces whereas in SAR imaging one deals with open surfaces. These two distinct cases of optical versus radar scattering are treated separately using two different reference frames; the '*Forward (anti-monostatic) Scattering Alignment (FSA)*' versus the '*Back Bistatic Scattering Alignment (BSA)*' from which the '*Monostatic Arrangement*' is derived as shown in Fig. 3.1. In the following, separately detailed for both the FSA and BSA systems are shown in Figs. 3.2 and 3.3.

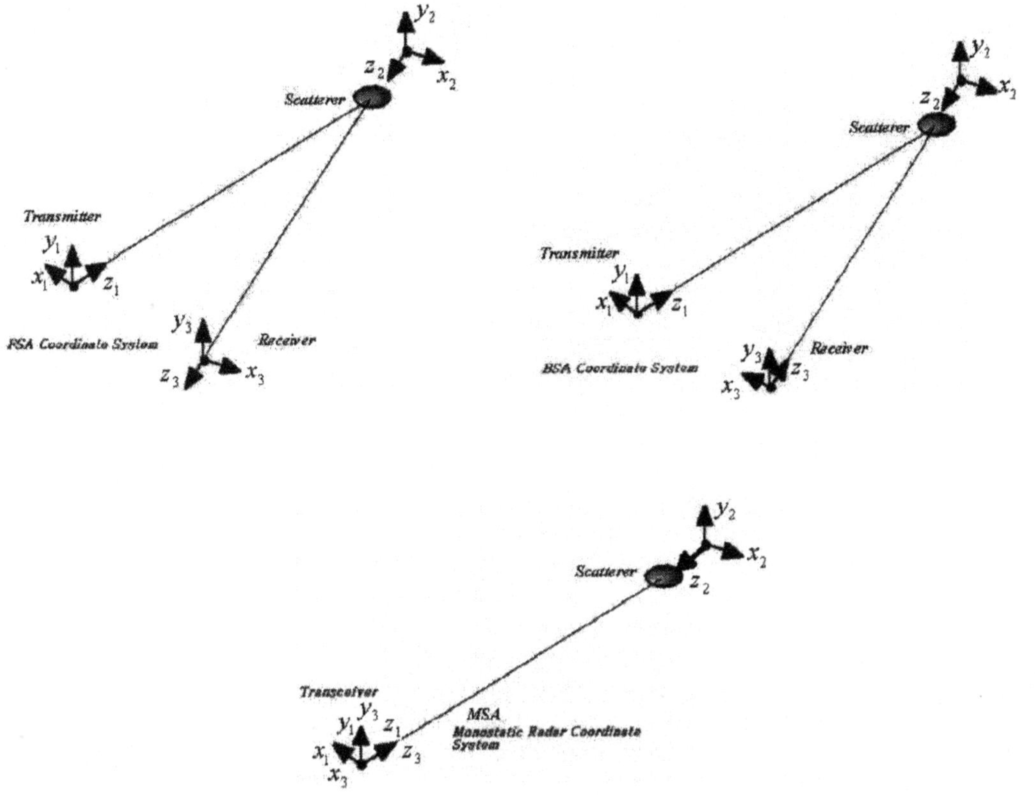

Fig. 3.1 Comparison of the FSA, BSA, and MSA Coordinate Systems

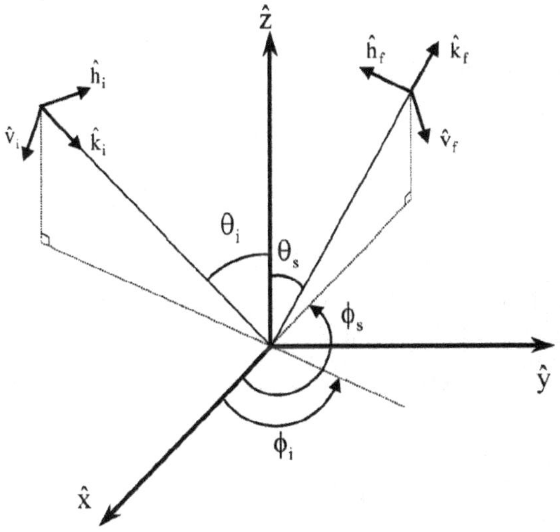

$$\hat{k}_f = \sin\theta_s \cos\phi_s \hat{x} + \sin\theta_s \sin\phi_s \hat{y} + \cos\theta_s \hat{z}$$
$$\hat{v}_f = \cos\theta_s \cos\phi_s \hat{x} + \cos\theta_s \sin\phi_s \hat{y} - \sin\theta_s \hat{z}$$
$$\hat{h}_f = -\sin\phi_s \hat{x} + \cos\phi_s \hat{y}$$

Fig. 3.2 Detailed Forward Scattering Alignment (FSA)

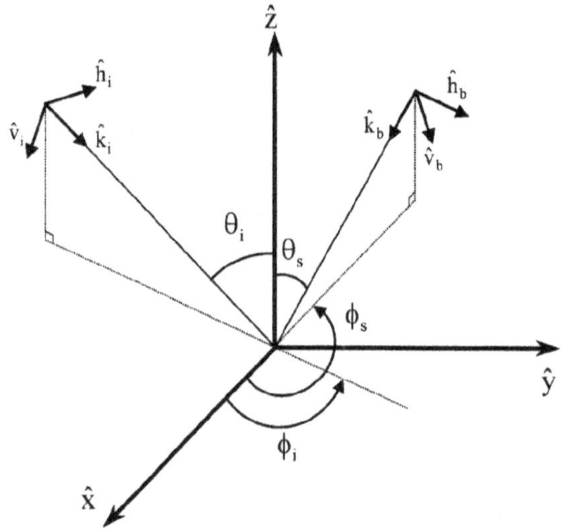

$$\hat{k}_b = -\sin\theta_s \cos\phi_s \hat{x} - \sin\theta_s \sin\phi_s \hat{y} - \cos\theta_s \hat{z}$$
$$\hat{v}_b = \cos\theta_s \cos\phi_s \hat{x} + \cos\theta_s \sin\phi_s \hat{y} - \sin\theta_s \hat{z}$$
$$\hat{h}_b = \sin\phi_s \hat{x} - \cos\phi_s \hat{y}$$

Fig. 3.3 Detailed Back Scattering Alignment (BSA)

3.2 The 2x2 Jones Forward [*J*] versus 2x2 Sinclair [*S*] Back-Scattering Matrices

Since we are dealing here with radar polarimetry, interferometry and polarimetric interferometry, the '*bistatic BSA reference frame*' is more relevant and is here introduced only for reasons of brevity but dealing both with the bistatic and the monostatic cases ; where we refer to [52, 53], [76] and [19] for a full treatment of the '*anti-monostatic FSA reference frame*'. Here, we refer to the dissertation of Papathanassiou [97], the textbook of Mott [76], and meticulous derivations of Lüneburg [52] for more detailed treatments of the subject matter, but we follow here the derivation presented in [19]. Using the coordinates of Fig. 3.1 with right-handed coordinate systems; $x_1\ y_1\ z_1$, $x_2\ y_2\ z_2$, $x_3\ y_3\ z_3$; denoting the transmitter, scatterer and receiver coordinates, respectively, a wave incident on the scatterer from the transmitter is given by the transverse components E_{x_1} and E_{y_1} in the right-handed coordinate system $x_1\ y_1\ z_1$ with the z_1 axis pointed at the target. The scatterer coordinate system $x_2\ y_2\ z_2$ is right-handed with z_2 pointing away from the scatterer toward a receiver. BSA Coordinate System $x_3\ y_3\ z_3$ is right-handed with z_3 pointing toward the scatterer. It would coincide with the transmitter system $x_1\ y_1\ z_1$ if the transmitter and receiver were co-located. The wave reflected by the target to the receiver may be described in either the transverse components E_{x_2} and E_{y_2} or by the reversed components E_{x_3} and E_{y_3}. Both conventions are used, leading to different matrix formulations. The incident and transmitted or reflected (scattered) fields are given by two-component vectors; therefore the relationship between them must be a 2x2 matrix. If the scattered field is expressed in $x_3\ y_3\ z_3$ coordinates (BSA), the fields are related by the Sinclair matrix $[S]$, thus

$$\begin{bmatrix} E^s_{x_3} \\ E^s_{y_3} \end{bmatrix} = \frac{1}{\sqrt{4\pi r_2}} \begin{bmatrix} S_{x_3x_1} & S_{x_3y_1} \\ S_{y_3x_1} & S_{y_3y_1} \end{bmatrix} \begin{bmatrix} E^i_{x_1} \\ E^i_{y_1} \end{bmatrix} e^{-jkr_2} \tag{3.4}$$

and if the scattered field is in $x_2\ y_2\ z_2$ coordinates (FSA), it is given by the product of the Jones matrix $[J]$ with the incident field, thus

$$\begin{bmatrix} E^S_{x_3} \\ E^S_{y_3} \end{bmatrix} = \frac{1}{\sqrt{4\pi r_2}} \begin{bmatrix} T_{x_2x_1} & T_{x_2y_1} \\ T_{y_2x_1} & T_{y_2y_1} \end{bmatrix} \begin{bmatrix} E^i_{x_1} \\ E^i_{y_1} \end{bmatrix} e^{-j k r_2} \tag{3.5}$$

In both equations the incident fields are those at the target, the received fields are measured at the receiver, and r_2 is the distance from target to receiver. The *'Sinclair matrix* $[S]$*'* is mostly used for **back-scattering**, but is readily extended to the **bistatic scattering** case. If the name **scattering matrix** is used without qualification, it normally refers to the Sinclair matrix $[S]$. In the general bistatic scattering case, the elements of the Sinclair matrix are not related to each other, except through the physics of the scatterer. However, if the receiver and transmitter are co-located, as in the **mono-static** or back-scattering situation, and if the medium between target and transmitter is reciprocal, mainly the Sinclair matrix $[S(AB)]$ is symmetric, i.e. $S_{AB} = S_{BA}$. The Jones matrix is used for the forward transmission case; and if the medium between target and transmitter, without Faraday rotation, the Jones matrix is usually normal. However, it should be noted that the Jones matrix is not in general normal, i.e., in general the Jones matrix does not have orthogonal eigenvectors. Even the case of only one eigenvector (and a generalized eigenvector) has been considered in optics (homogeneous and inhomogeneous Jones matrices). If the coordinate systems being used are kept in mind, the numerical subscripts can be dropped.

It is clear that in the bistatic case, the matrix elements for a target depend on the orientation of the target with respect to the line of sight from transmitter to target and on its orientation with respect to the target-receiver line of sight. In the forms (3.4) and (3.5) the matrices are **absolute matrices**, and with their use the phase of the scattered wave can be related to the phase of the transmitted wave, which is strictly required in the case of polarimetric interferometry. If this phase relationship is of no interest, as in the case of mono-static polarimetry, the distinct phase term can be neglected, and one of the matrix elements can be taken as real. The resulting form of the Sinclair matrix is called the **relative scattering matrix**. In general the elements of the scattering matrix are dependent on the frequency of the illuminating wave [19, 14, 15].

Another target matrix parameter that should be familiar to all who are interested in microwave remote sensing is the **radar cross section (RCS)**. It is proportional to the power received by a radar and is the area of an equivalent target that intercepts a power equal to its area multiplied by the power density of an incident wave and re-radiates it equally in all directions to yield a receiver power equal to that produced by the real target. The radar cross section depends on the polarization of both transmitting and receiving antennas. Thus the radar cross section may be specified as HH (horizontal receiving and transmitting antennas), HV (horizontal receiving and vertical transmitting antennas), etc. When considering ground reflections, the cross section is normalized by the size of the ground patch illuminated by the wave from the radar. The cross section is not sufficient to describe the polarimetric behavior of a target. In terms of the Sinclair matrix $[S]$, and the normalized effective lengths of transmitting and receiving antennas, $\hat{\mathbf{h}}_t$ and $\hat{\mathbf{h}}_r$, respectively, the radar cross section is

$$\sigma_{rt} = \left| \hat{h}_r^T \, [S] \, \hat{h}_t \right|^2 \tag{3.6}$$

A polarimetrically correct form of the **radar equation** that specifies received power in terms of antenna and target parameters is

$$W_{rt} = \frac{W_t G_t(\theta,\phi) A_{er}(\theta,\phi)}{(4\pi r_1 r_2)^2} \left| \hat{h}_r^T \, [S] \, \hat{h}_t \right|^2 \tag{3.7}$$

where W_t is the transmitter power and subscripts t and r identify transmitter and receiver, and its properties are defined in more detail in Mott [76] and in [19]. The effective antenna height $\hat{\mathbf{h}}(\theta,\phi)$, is defined via the electric field $\mathbf{E}^t(r,\theta,\phi)$, radiated by an antenna in its far field, as

$$\mathbf{E}^t(r,\theta,\phi) = \frac{jZ_0 I}{2\lambda r} \exp(-jkr)\hat{\mathbf{h}}(\theta,\phi), \tag{3.8}$$

with Z_0 the characteristic impedance, λ the wavelength, and I the antenna current.

3.3 <u>Basis Transformations of the 2x2 Sinclair Scattering Matrix [S]</u>

Redefining the incident and scattering cases in terms of the standard {H V} notation with $H = x$, $V = y$ and with proper re-normalization, we redefine (3.1) as

$$\mathbf{E}^s_{HV} = [S]_{HV} \mathbf{E}^*_{HV} \quad \text{or} \quad \mathbf{E}^s(HV) = [S(HV)]\mathbf{E}^*(HV) \tag{3.9}$$

where the complex conjugation results from inversion of the coordinate system in the BSA arrangement which invites a more rigorous formulation in terms of directional Sinclair vectors including the concepts of time reversal as treated by Lüneburg [52]. Using these Sinclair vector definitions one can show that the transformation from one orthogonal polarisation basis {H V} into another {i j} or {A B} is a unitary congruence (unitary consimilarity) transformation of the original Sinclair scattering matrix $[S]_{HV}$ into $[S]_{ij}$, where

$$[S]_{ij} = [U_2][S]_{HV}[U_2]^T \quad \text{or} \quad [S(ij)] = [U_2]\ [S(HV)] = [U_2]^T \tag{3.10}$$

with $[U_2]$ given by (2.23), so that the components of the general non-symmetric scattering matrix for the bistatic case in the new polarization basis, characterized by a complex polarization ratio ρ, can be written as [81, 25]

$$S_{ii} = \frac{1}{1+\rho\rho^*}\left[S_{HH} - \rho^* S_{HV} - \rho^* S_{VH} + \rho^{*2} S_{VV}\right] \tag{3.11}$$

$$S_{ij} = \frac{1}{1+\rho\rho^*}\left[\rho S_{HH} + S_{HV} - \rho\rho^* S_{VH} - \rho^* S_{VV}\right]$$

$$S_{ji} = \frac{1}{1+\rho\rho^*}\left[\rho S_{HH} - \rho\rho^* S_{HV} + S_{VH} - \rho^* S_{VV}\right]$$

$$S_{jj} = \frac{1}{1+\rho\rho^*}\left[\rho^2 S_{HH} + \rho S_{HV} - \rho S_{VH} + S_{VV}\right]$$

There exist three invariants for the general bistatic case (BSA) under the change-of-basis transformation as given by (3.5):

$$\text{(i) } \kappa_4 = Span[S] = \left\{|S_{HH}|^2 + |S_{HV}|^2 + |S_{VH}|^2 + |S_{VV}|^2\right\} = \left\{|S_{ii}|^2 + |S_{ij}|^2 + |S_{ji}|^2 + |S_{jj}|^2\right\} \tag{3.12}$$

confirms that the total power is conserved, and it is known as Kennaugh's span-invariant κ_4;

(ii) $S_{HV} - S_{VH} = S_{ij} - S_{ji}$, for monostatic case (3.13)

warranting symmetry of the scattering matrix in any polarization basis as long as the BSA for the strictly mono-static but not general bistatic case is implied;

(iii) $Det\{[S]_{HV}\} = Det\{[S]_{ij}\}$ or $Det\{[S(HV)]\} = Det\{[S(ij)]\}$ (3.14)

due to the fact that $Det\{[U_2]\} = 1$ implies determinantal invariance.

In addition, diagonalization of the scattering matrix, also for the general bistatic case, can always be obtained but requires mixed basis representations by using the *'Singular Value Decomposition Theorem (SVD)'* [52, 53] so that the diagonalized scattering matrix $[S_D]$ can be obtained by the left and right singular vectors, where

$$[S_D] = [Q_L][S][Q_R] \text{ with } [S_D] = \begin{bmatrix} S_{\lambda_1} & 0 \\ 0 & S_{\lambda_2} \end{bmatrix} \tag{3.15}$$

and $|Det\{[Q_L]\}| = |Det\{[Q_R]\}| = 1$

and S_{λ_1} and S_{λ_2} denote the diagonal eigenvalues of $[S]$, and the diagonal elements S_{λ_1} and S_{λ_2} can be taken as real nonnegative and are known as the singular values of the matrix $[S]$. For the symmetric scattering matrices in the mono-static case (MSA), diagonalization is achieved according to the *unitary consimilarity* transform for which

$$[Q_R] = [Q_L]^T \tag{3.16}$$

and above equations will simplify due to the restriction of symmetry $S_{ij} = S_{ji}$.

3.4 The 4x4 Mueller (Forward Scattering) [*M*] and the 4x4 Kennaugh (Back-Scattering) [*K*] Power Density Matrices

For the partially polarized cases, there exists an alternate formulation of expressing the scattered wave in terms of the incident wave via the 4x4 Mueller$[M]$ and Kennaugh $[K]$ matrices for the FSA and BSA coordinate formulations, respectively, where

$$[q^s] = [M][q^i] \tag{3.17}$$

For the purely coherent case, $[M]$ can formally be related to the coherent Jones Scattering Matrix $[T]$ as

$$[M] = \begin{bmatrix} 1 & 1 & 1 & -1 \end{bmatrix} [A]^{T-1} ([T] \otimes [T]^*) [A]^{-1} = [A]([T] \otimes [T]^*)[A]^{-1} \tag{3.18}$$

with $\otimes$ symbolizing the standard Kronecker tensorial matrix product relations [112] provided in (A.1), Appendix A, and the 4x4 expansion matrix $[A]$ is given by [76] as

$$[A] = \begin{bmatrix} 1 & 0 & 0 & 1 \\ 1 & 0 & 0 & -1 \\ 0 & 1 & 1 & 0 \\ 0 & j & -j & 0 \end{bmatrix} \tag{3.19}$$

with the elements M_{ij} of $[M]$, given in (B.1), Appendix B.

Specifically we find that if $[T]$ is normal, i.e. $[T][T]^{T*} = [T]^{T*}[T]$, then $[M]$ is also normal, i.e. $[M][M]^T = [M]^T[M]$.

Similarly, for the purely coherent case [76], $[K]$ can formally be related to the *coherent Sinclair matrix* $[S]$ as

$$[K] = 2[A]^{T-1}([S] \otimes [S]^*)[A]^{-1} \tag{3.20}$$

where

$$[A]^{T-1} = \frac{1}{2}[A]^* \tag{3.21}$$

and for a symmetric Sinclair matrix $[S]$, then $[K]$ is symmetric, keeping in mind the '*mathematical formalism*' $[M] = diag[1 \quad 1 \quad 1 \quad -1][K]$, but great care must be taken in strictly distinguishing the physical meaning of $[K]$ versus $[M]$ in terms of $[S]$ versus $[T]$ respectively. Thus, if $[S]$ is symmetric, $S_{HV} = S_{VH}$, then $[K]$ is symmetric, $K_{ij} = K_{ji}$; and the correct elements for $[M]$, $[K]$ and the symmetric cases are presented in (B.1 – B.5), Appendix B.

3.5 The 2x2 Graves Polarization Power Scattering Matrix [*G*]

Kennaugh introduces, next to the Kennaugh matrix $[K]$, another formulation $[G]$, for expressing the power in the scattered wave $\mathbf{E}^S$ to the incident wave $\mathbf{E}^i$ for the coherent case in terms of the so-called '*Graves polarization coherent power scattering matrix* $[G]$', where

$$P^S = \frac{1}{8\pi\, Z_0 r_2^2} \mathrm{E}^{iT*}[G]\mathrm{E}^i \tag{3.22}$$

so that in terms of the Kennaugh elements K_{ij}, defined in the appendix, for the mono-static case

$$[G] = \langle [S]^{T*}[S] \rangle = \begin{bmatrix} K_{11} + K_{12} & K_{13} - jK_{14} \\ K_{13} + jK_{14} & K_{11} - K_{12} \end{bmatrix} \tag{3.23}$$

By using a single coordinate system for (x_1, y_1, z_1) and (x_3, y_3, z_3) for the monostatic case, as in Fig. 3.1, and also described in detail in [19], it can be shown that for a scatterer ensemble (e.g. precipitation) for which individual scatterers move slowly compared to a period of the illuminating wave, and quickly compared to the time-averaging of the receiver, time-averaging can be adjusted to find the decomposed power scattering matrix $\langle [G] \rangle$, as

$$\langle [G] \rangle = \langle [S(t)]^{T*}[S(t)] \rangle = [G_H] + [G_V] = \left\{ \left\langle \begin{bmatrix} |S_{HH}|^2 & S_{HH}^* S_{HV} \\ S_{HH} S_{HV}^* & |S_{HV}|^2 \end{bmatrix} \right\rangle + \left\langle \begin{bmatrix} |S_{VH}|^2 & S_{VH}^* S_{VV} \\ S_{VH} S_{VV}^* & |S_{VV}|^2 \end{bmatrix} \right\rangle \right\} \tag{3.24}$$

This shows that the time averaged *'Graves Power Scattering Matrix* $\langle [G] \rangle$ ', first introduced by Kennaugh [4, 5], can be used to divide the powers that are received by linear horizontally and vertically by polarized antennas, as discussed in more detail in [19] and in [113]. It should be noted that a similar decomposition also exists for the Muller/Jones matrices, commonly denoted as FSA power scattering matrix

$$\langle [F] \rangle = \langle [T(t)]^{+} [T(t)] \rangle = [F_H] + [F_V] = \left\{ \left\langle \begin{bmatrix} |T_{HH}|^2 & T_{HH}^{*} T_{HV} \\ T_{HH} T_{HV}^{*} & |T_{HV}|^2 \end{bmatrix} \right\rangle + \left\langle \begin{bmatrix} |T_{VH}|^2 & T_{VH}^{*} T_{VV} \\ T_{VH} T_{VV}^{*} & |T_{VV}|^2 \end{bmatrix} \right\rangle \right\} \tag{3.25}$$

which is not further analyzed here [113].

3.6 Co/Cross-Polar Backscattering Power Decomposition for the One-Antenna (Transceiver) and the Matched Two-Antenna (Quasi-Monostatic) Cases

Assuming that the scatterer is placed in free unbounded space and that no polarization state transformation occurs along the propagation path from the transmitter (T) to the scatterer incidence (S), and along that from the scatterer(s) to the receiver (R), then the value of the terminal voltage of the receiver, V_R, induced by an arbitrarily scattered wave $\mathbf{E}_R$ at the receiver, is defined by the r*adar brightness function* V_R, and the corresponding received power P_R expression

$$V_R = \hat{\mathbf{h}}_{\mathbf{R}}^T \mathbf{E}_{\mathbf{R}} \qquad P_R = \frac{1}{2} V_R^{*} V_R \tag{3.26}$$

with the definition of the Kennaugh matrix $[K]$ in terms of the Sinclair matrix $[S]$, the received power or *radar brightness* function may be re-expressed

$$P_{RT} = \frac{1}{2} \left| \hat{\mathbf{h}}_R \, [\mathrm{S}] \, \mathbf{E}_T \right|^2 = \frac{1}{2} \mathbf{q}_R^T \, [K] \, \mathbf{q}_T \tag{3.27}$$

where $\mathbf{q}_R$ and $\mathbf{q}_T$ the corresponding normalized Stokes' vectors.

For the one-antenna (transceiver) case the co-polar channel (c) and the cross-polar channel (x) powers become:

$$P_c = \frac{1}{2} \left| \hat{\mathbf{h}}_T^{\;T} \, [\mathrm{S}] \, \mathbf{E}_T \right|^2 = \frac{1}{2} \mathbf{q}_T^{\;T} \, [K_c] \, \mathbf{q}_T \tag{3.28}$$

$$P_x = \frac{1}{2} \left| \hat{\mathbf{h}}_{T_\perp}^{\;T} \, [\mathrm{S}] \, \mathrm{E}_T \right|^2 = \frac{1}{2} \mathbf{q}_T^{\;T} \, [K_x] \, \mathbf{q}_T \tag{3.29}$$

with

$$[K_c] = \left([A]^{-1}\right)^T ([T] \otimes [T]^{*}) [A]^{-1} = [C][K] \tag{3.30}$$

$$[K_x] = \left([A]^{-1}\right)^T ([Y][T] \otimes [T]^{*}) [A]^{-1} = [C_x][K] \tag{3.31}$$

and

$$[C]=\begin{bmatrix}1&0&0&0\\0&1&0&0\\0&0&1&0\\0&0&0&-1\end{bmatrix}\quad [Y]=\begin{bmatrix}0&0&0&1\\0&-1&0&0\\0&0&-1&0\\1&0&0&0\end{bmatrix}\quad [X]=\begin{bmatrix}1&0&0&0\\0&-1&0&0\\0&0&-1&0\\0&0&0&1\end{bmatrix} \tag{3.32}$$

For the *Two-Antenna Dual Polarization* case, in which one antenna serves as a transmitter and the other as the receiver, the optimal received power P_m for the *'matched case'* becomes by using the matching condition

$$\hat{\mathbf{h}}_{\mathbf{R}_m} = \mathbf{E}_s^* / \|\mathbf{E}_s\| \tag{3.33}$$

so that

$$P_m = \mathbf{q}_T^T [K_m] \mathbf{q}_T, \text{ where } [K_m]=[K_c]+[K_x]=[K_{11}][K], \text{ and } [K_{11}]=\begin{bmatrix}1&0&0&0\\0&0&0&0\\0&0&0&0\\0&0&0&0\end{bmatrix} \tag{3.34}$$

which represent an essential relationship for determining the optimal polarization states from the optimization of the Kennaugh matrix.

3.7 The Scattering Feature Vectors : The Lexicographic and the Pauli Feature Vectors

Up to now we have introduced three descriptions of the scattering processes in terms of the 2x2 Jones versus Sinclair, $[T]$ versus $[S]$, the 2x2 power scattering matrices, $[F]$ versus $[G]$, and the 4x4 power density Muller versus Kennaugh matrices, $[M]$ versus $[K]$. Alternatively, the polarimetric scattering problem can be addressed in terms of a vectorial feature descriptive formulation [114] borrowed from vector signal estimation theory. This approach replaces the 2x2 scattering matrices $[T]$ versus $[S]$, the 2x2 power scattering matrices $[F]$ versus $[G]$, and the 4x4 Muller $[M]$ versus Kennaugh $[K]$ matrices by an equivalent four-dimensional complex scattering feature vector $\mathbf{f}_4$, formally defined for the general bi-static case as

$$[S]_{HV} = \begin{bmatrix} S_{HH} & S_{HV} \\ S_{VH} & S_{VV} \end{bmatrix} \Rightarrow \mathbf{f}_4 = F\{[S]\} = \frac{1}{2} Trace\{[S]\ \psi\} = [f_0 \quad f_1 \quad f_2 \quad f_3]^T \tag{3.35}$$

where $F\{[S]\}$ is the matrix vectorization operator $Trace\{[S]\}$ is the sum of the diagonal elements of $[S]$, and ψ is a complete set of 2x2 complex basis matrices under a hermitian inner product. For the vectorization of any complete orthonormal basis set [97] of four 2x2 matrices that leave the (Euclidean) norm of the scattering feature vector invariant, can be used, and there are two such bases favored in the polarimetric radar literature; one being the *'lexicographic basis'* $[\Psi_L]$, and the other *'Pauli spin matrix set'* $[\Psi_P]$. We note here that the distinction between the lexicographic and Pauli-based feature vector representation is related to Principal and Independent Component Analysis (PCA/ICA) which is an interesting topic for future research.

(i) The *'Lexicographic Feature vector'*: $\mathbf{f}_{4L}$, is obtained from the simple lexicographic expansion of $[S]$ using $[\Psi_L]$, with

$$[\Psi_L] \equiv \left\{ 2\begin{bmatrix} 1 & 0 \\ 0 & 0 \end{bmatrix} \; 2\begin{bmatrix} 0 & 1 \\ 0 & 0 \end{bmatrix} \; 2\begin{bmatrix} 0 & 0 \\ 1 & 0 \end{bmatrix} \; 2\begin{bmatrix} 0 & 0 \\ 0 & 1 \end{bmatrix} \right\} \tag{3.36}$$

so that the corresponding feature vector becomes

$$\mathbf{f}_{4L} = [S_{HH} \quad S_{HV} \quad S_{VH} \quad S_{VV}]^T \tag{3.37}$$

(ii) The Pauli Feature vector $\mathbf{f}_{4P}$ is obtained from the renowned complex Pauli spin matrix basis set $[\Psi_P]$ which in a properly re-normalized presentation is here defined as

$$[\Psi_P] \equiv \left\{ \sqrt{2}\begin{bmatrix} 1 & 0 \\ 0 & 1 \end{bmatrix} \; \sqrt{2}\begin{bmatrix} 1 & 0 \\ 0 & -1 \end{bmatrix} \; \sqrt{2}\begin{bmatrix} 0 & 1 \\ 1 & 0 \end{bmatrix} \; \sqrt{2}\begin{bmatrix} 0 & -j \\ j & 0 \end{bmatrix} \right\} \tag{3.38}$$

resulting in the *'polarimetric correlation phase'* preserving *'Pauli Feature Vector'*.

$$\mathbf{f}_{4P} = [f_0 \quad f_1 \quad f_2 \quad f_3]_P^T = \frac{1}{\sqrt{2}} [S_{HH} + S_{VV} \quad S_{VV} - S_{HH} \quad S_{HV} + S_{VH} \quad j(S_{HV} - S_{VH})]^T \tag{3.39}$$

where the corresponding scattering matrix $[S]_P$ is related to the $\mathbf{f}_{4P} = [f_0 \quad f_1 \quad f_2 \quad f_3]_P^T$ by

$$[S]_P = \frac{1}{\sqrt{2}} \begin{bmatrix} f_0 - f_1 & f_2 - jf_3 \\ f_2 + jf_1 & f_0 + f_1 \end{bmatrix} = [S] \tag{3.40}$$

3.8 The Unitary Transformations of the Feature Vectors

The insertion of the factor 2 in (3.36) versus the factor $\sqrt{2}$ in (3.38) arises from the *'total power invariance'*, i.e. keeping the norm independent from the choice of the basis matrices Ψ, so that

$$\|\mathbf{f}_4\| = \mathbf{f}_4^\dagger \cdot \mathbf{f}_4 = \frac{1}{2}\,\mathrm{Span}\{[S]\} = \frac{1}{2} Trace\{[S][S]^\dagger\} = \frac{1}{2}\left(|S_{HH}|^2 + |S_{HV}|^2 + |S_{VH}|^2 + |S_{VV}|^2\right) = \kappa_4 \tag{3.41}$$

This constraint forces the transformation from the lexicographic to the Pauli-based feature vector [52, 53, 114], or to any other desirable one, to be unitary, where with

$$\mathbf{f}_{4P} = [D_4]\mathbf{f}_{4L} \qquad \text{and reversely} \qquad \mathbf{f}_{4L} = [D_4]^{-1}\mathbf{f}_{4P} \tag{3.42}$$

we find

$$[D_4] = \frac{1}{\sqrt{2}} \begin{bmatrix} 1 & 0 & 0 & 1 \\ 1 & 0 & 0 & -1 \\ 0 & 1 & 1 & 0 \\ 0 & j & -j & 0 \end{bmatrix} \qquad [D_4]^{-1} = [D_4]^\dagger = \frac{1}{\sqrt{2}} \begin{bmatrix} 1 & 1 & 0 & 0 \\ 0 & 0 & 1 & -j \\ 0 & 0 & 1 & j \\ 1 & -1 & 0 & 0 \end{bmatrix} \tag{3.43}$$

Furthermore, these special unitary matrices relating the feature vectors control the more general cases of transformations related to the change of polarization basis. By employing the Kronecker direct tensorial product of matrices, symbolized by $\otimes$, we obtain, the transformation for the scattering vector from the linear $\{\hat{\mathbf{u}}_H, \hat{\mathbf{u}}_V\}$ to any other elliptical polarization basis $\{\hat{\mathbf{u}}_A, \hat{\mathbf{u}}_B\}$, characterized by the complex polarization ratio by

$$\mathbf{f}_{4L}(AB) = [U_{4L}]\mathbf{f}_{4L}(HV) \quad \text{and} \quad \mathbf{f}_{4P}(AB) = [U_{4P}]\mathbf{f}_{4P}(HV) \tag{3.44}$$

where $[U_{4L}]$ is the transformation matrix for the conventional feature vector $\mathbf{f}_{4L}$

Here we note that in order to obtain the expression $[U_{4L}] = [U_2] \otimes [U_2]^T$, the unitary congruence (unitary consimilarity) transformation for the Sinclair scattering matrix in the reciprocal case was used. This implies however that we must distinguish between forward scattering and backscattering (and so also bistatic scattering); where for the reciprocal backscatter case the 3-dimensional target feature vectors ought to be used. These features lead to interesting questions which however need more in depth analyses for which the ubiquity of the Time Reversal operation shows up again.

$$[U_{4L}] = [U_2] \otimes [U_2]^T = \frac{1}{1+\rho\rho^*} \begin{bmatrix} 1 & -\rho^* & -\rho^* & \rho^{*2} \\ \rho & 1 & -\rho\rho^* & -\rho^* \\ \rho & -\rho\rho^* & 1 & -\rho^* \\ \rho^2 & \rho & \rho & 1 \end{bmatrix} \tag{3.45}$$

and $[U_{4P}]$ is the homologous transformation matrix for the Pauli-based feature vector $\mathbf{f}_{4P}$

$$[U_{4P}] = [D_4][U_{4L}][D_4]^\dagger \tag{3.46}$$

where $[U_{4L}]$ and $[U_{4P}]$ are special 4x4 unitary matrices for which with $[I_4]$ denoting the 4x4 identity matrix

$$[U_4][U_4] = [I_4] \quad \text{and} \quad Det\{[U_4]\} = 1 \tag{3.47}$$

Kennaugh matrices and covariance matrices are based on completely different concepts (notwithstanding their formal relationships) and must be clearly separated which is another topic for future research.

The main advantage of using the scattering feature vector, $\mathbf{f}_{4L}$ or $\mathbf{f}_{4P}$, instead of the Sinclair scattering matrix $[S]$ and the Kennaugh matrix $[K]$, is that it enables the introduction of the covariance matrix decomposition for partial scatterers of a dynamic scattering environment. However, there does not exist a physical but only a strict relationship mathematical between the two alternate concepts for treating the partially coherent case, which is established and needs always to be kept in focus [114]. It should be noted that besides the covariance matrices the so-called (normalized) correlation matrices are often used advantageously especially when the eigenvalues of a covariance matrix have large variations.

3.9 The Polarimetric Covariance Matrix

In most radar applications, the scatterers are situated in a dynamically changing environment and are subject to spatial (different view angles as in 'SAR') and temporal variations (different hydro-meteoric states in 'RAD-MET'), if when illuminated by a monochromatic waves cause the back-scattered wave to be partially polarized with incoherent scattering contributions so that "$\langle [S] \rangle = \langle [S(\mathbf{r},t)] \rangle$". Such scatterers, analogous to the partially polarized waves are called partial scatterers [78, 90]. Whereas the Stokes vector, the wave coherency matrix, and the Kennaugh/Mueller matrix representations provide a first approach of dealing with partial scattering descriptions, the unitary matrix derived from the scattering feature $\mathbf{f}_4$ vector provides another approach borrowed from decision and estimation signal theory [115] which are currently introduced in Polarimetric SAR and Polarimetric-Interferometric SAR analyses, and these need to be introduced here. However, even if the environment is dynamically changing one has to make assumption concerning stationarity (at least over timescales of interest), homogeneity and ergodicity. This can be analyzed more precise by introducing the concept of space and time varying stochastic processes.

The 4x4 lexicographic polarimetric covariance matrix $[C_{4L}]$ and the Pauli-based covariance matrix $[C_{4P}]$ are defined, using the outer product $\otimes$.of the feature vector with its conjugate transpose as

$$[C_{4L}] = \langle \mathbf{f}_{4L} \cdot \mathbf{f}_{4L}^{\dagger} \rangle \quad \text{and} \quad [C_{4P}] = \langle \mathbf{f}_{4P} \cdot \mathbf{f}_{4P}^{\dagger} \rangle \tag{3.48}$$

where $\langle \ldots \rangle$ indicates temporal or spatial ensemble averaging, assuming homogeneity of the random medium. The lexicographic covariance matrix $[C_4]$ contains the complete information in amplitude and phase variance and correlation for all complex elements of $[S]$ with

$$[C_{4L}] = \langle \mathbf{f}_{4L} \cdot \mathbf{f}_{4L}^{\dagger} \rangle = \begin{bmatrix} \langle |S_{HH}|^2 \rangle & \langle S_{HH} S_{HV}^* \rangle & \langle S_{HH} S_{VH}^* \rangle & \langle S_{HH} S_{VV}^* \rangle \\ \langle S_{HV} S_{HH}^* \rangle & \langle |S_{HV}|^2 \rangle & \langle S_{HV} S_{VH}^* \rangle & \langle S_{HV} S_{VV}^* \rangle \\ \langle S_{VH} S_{HH}^* \rangle & \langle S_{VH} S_{HV}^* \rangle & \langle |S_{VH}|^2 \rangle & \langle S_{VH} S_{VV}^* \rangle \\ \langle S_{VV} S_{HH}^* \rangle & \langle S_{VV} S_{HV}^* \rangle & \langle S_{VV} S_{VH}^* \rangle & \langle |S_{VV}|^2 \rangle \end{bmatrix} \tag{3.49}$$

and

$$[C_{4P}] = \langle \mathbf{f}_{4P} \cdot \mathbf{f}_{4P}^{\dagger} \rangle = \langle [D_4]\mathbf{f}_{4L} \cdot \mathbf{f}_{4L}^{\dagger} [D_4]^{\dagger} \rangle = [D_4] \langle \mathbf{f}_{4L} \cdot \mathbf{f}_{4L}^{\dagger} \rangle [D_4]^{\dagger} = [D_4][C_{4L}][D_4]^{\dagger} \tag{3.50}$$

Both the *'Lexicographic Covariance* $[C_{4L}]$ ' and the *'Pauli-based Covariance* $[C_{4P}]$*]'* matrices are hermitian positive semi-definite matrices which implies that these possess real non-negative eigenvalues and orthogonal eigenvectors. Incidentally, those can be mathematically related directly to the Kennaugh matrix $[K]$, which is not shown here; however, there does not exist a physical relationship between the two presentations which must always be kept in focus.

The transition of the covariance matrix from the particular linear polarization reference basis {H V} into another elliptical basis {A B}, using the change-of-basis transformations defined in (3.41 – 3.45), where for

$$[C_{4L}(AB)] = \langle \mathbf{f}_{4L}(AB) \cdot \mathbf{f}_{4L}^{\dagger}(AB) \rangle = [U_4] \langle \mathbf{f}_{4L}(HV) \cdot \mathbf{f}_{4L}^{\dagger}(HV) \rangle [U_4]^{\dagger} = [D_4][C_{4L}(HV)][D_4]^{\dagger} \tag{3.51}$$

and for

$$[C_{4P}(AB)] = \langle \mathbf{f}_{4P}(AB) \cdot \mathbf{f}_{4P}^{\dagger}(AB) \rangle = [U_4] \langle \mathbf{f}_{4P}(HV) \cdot \mathbf{f}_{4P}^{\dagger}(HV) \rangle [U_4]^{\dagger} = [D_4][C_{4P}(HV)][D_4]^{\dagger} \tag{3.52}$$

The lexicographic and Pauli-based covariance matrices, $[C_{4L}]$ and $[C_{4P}]$, contain, in the most general case, as defined in (3.49) and (3.50), sixteen independent parameters, namely four real power densities and six complex phase correlation parameters.

3.10 The Monostatic Reciprocal Back-Scattering Cases

For a reciprocal target matrix, in the mono-static (backscattering) case, the reciprocity constrains the Jones matrix to be usually normal, and the Sinclair scattering matrix to be symmetrical, i.e. $S_{HV} = S_{VH}$, which further reduces the expressions of $[G]$ and $[K]$. Furthermore, the four-dimensional scattering feature vector $\mathbf{f}_4$ reduces to a three- dimensional scattering feature vector $\mathbf{f}_3$ such that following [97]

$$\mathbf{f}_{3L} = [Q] \text{ , } \mathbf{f}_{4L} = \begin{bmatrix} S_{HH} & \sqrt{2}S_{HV} & S_{VV} \end{bmatrix}^T \text{ , } S_{HV} = S_{VH} \tag{3.53}$$

where with $[I_3]$ being the unit 3x3 matrix and always keeping in mind that the transformation between lexicographic and Pauli ordering is a direct transformation of the scattering matrix (and not only of the covariance matrices)

$$[Q] = \begin{bmatrix} 1 & 0 & 0 & 0 \\ 0 & \frac{1}{\sqrt{2}} & \frac{1}{\sqrt{2}} & 0 \\ 0 & 0 & 0 & 1 \end{bmatrix} \quad \text{and} \quad [Q][Q]^T = [I_3] \tag{3.54}$$

and the factor $\sqrt{2}$ needs to be retained in order to keep the vector norm consistent with the span invariance κ.

Similarly, the reduced Pauli feature vector $\mathbf{f}_{3P}$ becomes

$$\mathbf{f}_{3P} = [Q] \text{ , } \mathbf{f}_{4P} = \frac{1}{\sqrt{2}} \begin{bmatrix} S_{HH} + S_{VV} & S_{HH} - S_{VV} & 2S_{HV} \end{bmatrix}^T \text{ , } S_{HV} = S_{VH} \tag{3.55}$$

The three-dimensional scattering feature vector from the lexicographic to the Pauli-based matrix basis, and vice versa, are related as

$$\mathbf{f}_{3P} = [D_3]\mathbf{f}_{3L} \quad \text{and} \quad \mathbf{f}_{3L} = [D_3]^{-1}\mathbf{f}_{3P} \tag{3.56}$$

with $[D_3]$ defining a special 3x3 unitary matrix

$$[D_3] = \frac{1}{\sqrt{2}} \begin{bmatrix} 1 & 0 & 1 \\ 1 & 0 & -1 \\ 0 & \sqrt{2} & 0 \end{bmatrix} \quad \text{and} \quad [D_3]^{-1} = [D_3]^{\dagger} = \frac{1}{2} \begin{bmatrix} 1 & 1 & 0 \\ 0 & 0 & \sqrt{2} \\ 1 & -1 & 0 \end{bmatrix} \tag{3.57}$$

The change-of-basis transformation for the reduced scattering vectors in terms of the complex polarization ratio ρ of the new basis is given by

$$\mathbf{f}_{3L}(AB) = [U_{3L}(\rho)]\mathbf{f}_{3L}(HV) \quad \text{and} \quad \mathbf{f}_{3P}(AB) = [U_{3P}(\rho)]\mathbf{f}_{3P}(HV) \tag{3.58}$$

where

$$[U_{3L}] = \frac{1}{1+\rho\rho^*} \begin{bmatrix} 1 & \sqrt{2}\rho & \rho^2 \\ -\sqrt{2}\rho^* & 1-\rho\rho^* & \sqrt{2}\rho \\ \rho^{*2} & -\sqrt{2}\rho^* & 1 \end{bmatrix} \tag{3.59}$$

and

$$[U_{3P}]=[D_3][U_{3L}][D_3]^\dagger=\frac{1}{2(1+\rho\rho^*)}\begin{bmatrix}2+\rho^2+\rho^{*^2} & \rho^{*^2}-\rho^2 & 2(\rho-\rho^*)\\ \rho^2-\rho^{*^2} & 2-(\rho^2+\rho^{*^2}) & 2(\rho+\rho^*)\\ 2(\rho-\rho^*) & -2(\rho+\rho^*) & 2(1-\rho\rho^*)\end{bmatrix} \tag{3.60}$$

which are 3x3 special unitary matrices.

Thus, a reciprocal scatterer is completely described either by the 3x3 *'Polarimetric Covariance Matrix* $[C_{3L}]$'

$$[C_{3L}]=\langle \mathbf{f}_{3L}\cdot\mathbf{f}_{3L}^\dagger\rangle=\begin{bmatrix}\langle |S_{HH}|^2\rangle & \sqrt{2}\langle S_{HH}S_{HV}^*\rangle & \langle S_{HH}S_{VV}^*\rangle\\ \sqrt{2}\langle S_{HV}S_{HH}^*\rangle & 2\langle |S_{HV}|^2\rangle & \sqrt{2}\langle S_{HV}S_{VV}^*\rangle\\ \langle S_{VV}S_{HH}^*\rangle & \sqrt{2}\langle S_{VV}S_{HV}^*\rangle & \langle |S_{VV}|^2\rangle\end{bmatrix} \tag{3.61}$$

or by the 3x3 *'Polarimetric Pauli Coherency Matrix* $[C_{3P}]$'

$$[C_{3P}]=\langle \mathbf{f}_{3P}\cdot\mathbf{f}_{3P}^\dagger\rangle=\frac{1}{2}\begin{bmatrix}\langle |S_{HH}+S_{VV}|^2\rangle & \langle (S_{HH}+S_{VV})(S_{HH}-S_{VV})^*\rangle & 2\langle (S_{HH}+S_{VV})S_{HV}^*\rangle\\ \langle (S_{HH}-S_{VV})(S_{HH}+S_{VV})^*\rangle & \langle |S_{HH}-S_{VV}|^2\rangle & 2\langle (S_{HH}-S_{VV})S_{HV}^*\rangle\\ 2\langle S_{HV}(S_{HH}+S_{VV})^*\rangle & 2\langle S_{HV}(S_{HH}-S_{VV})^*\rangle & 4\langle |S_{HV}|^2\rangle\end{bmatrix} \tag{3.62}$$

where the relation between the 3x3 Pauli coherency matrix $[C_{3P}]$ and the 3x3 covariance matrix $[C_{3L}]$ is given by

$$[C_{3L}]=\langle \mathbf{f}_{3L}\cdot\mathbf{f}_{3L}^\dagger\rangle=\langle [D_3]\mathbf{f}_{3P}\cdot\mathbf{f}_{3P}^\dagger[D_3]^\dagger\rangle=[D_3]\langle \mathbf{f}_{3P}\cdot\mathbf{f}_{3P}^\dagger\rangle[D_3]^\dagger=[D_3][C_{3P}][D_3]^\dagger \tag{3.63}$$

and

$$[C_{3L}(AB)]=\langle \mathbf{f}_{3L}(AB)\cdot\mathbf{f}_{3L}^\dagger(AB)\rangle=[U_{3L}]\langle \mathbf{f}_{3L}(HV)\cdot\mathbf{f}_{3L}^\dagger(HV)\rangle[U_{3L}]^\dagger=[U_{3L}][C_{3L}(HV)][U_{3L}]^\dagger \tag{3.64}$$

$$[C_{3P}(AB)]=\langle \mathbf{f}_{3P}(AB)\cdot\mathbf{f}_{3P}^\dagger(AB)\rangle=[U_{3P}]\langle \mathbf{f}_{3P}(HV)\cdot\mathbf{f}_{3P}^\dagger(HV)\rangle[U_{3P}]^\dagger=[U_{3P}][C_{3P}(HV)][U_{3P}]^\dagger \tag{3.65}$$

where

$$[U_{3L}(\rho)][U_{3L}(\rho)]^+=[I_3] \quad \text{and} \quad Det\{[U_{3L}(\rho)]\}=1 \tag{3.66}$$

and

$$\|f_{3L}\|^2=\|f_{3P}\|^2=\frac{1}{2}\,Span\,\{[S]\}=\frac{1}{2}\,Trace\{[S][S]^\dagger\}=\frac{1}{2}\,\{|S_{HH}|^2+2|S_{HV}|^2+|S_{VV}|^2\}=\kappa_3 \tag{3.67}$$

3.11 Co/Cross-polar Power Density and Phase Correlation Representations

The Covariance matrix elements are directly related to polarimetric radar measurables, comprised of the *Co/Cross-Polar Power Densities* $P_c(\rho)$, $P_x(\rho)$, $P_c^{\perp}(\rho)$, and the *Co/Cross-Polar Phase Correlations* $R_c(\rho)$, $R_x(\rho)$, $R_x^{\perp}(\rho)$, [81], *where*

$$[C_{3L}(\rho)] = \begin{bmatrix} P_c(\rho) & \sqrt{2}\,R_x(\rho) & R_c(\rho) \\ \sqrt{2}\,R_x(\rho)^* & 2P_x(\rho) & \sqrt{2}\,R_x^{\perp}(\rho)^* \\ R_c(\rho)^* & \sqrt{2}\,R_x^{\perp}(\rho) & P_c^{\perp}(\rho) \end{bmatrix} \tag{3.68}$$

Once the covariance matrix has been measured in one basis, e.g., $[C_{3L}(H,V)]$ in {H V} basis, it can easily be determined analytically for any other basis by definition of (3.60). Plotting the mean power returns and phase correlations as function of the complex polarization ratio ρ or the geometrical polarization ellipse parameters ψ, χ, of (3), yields the familiar '*polarimetric signature plots*'. In addition, the expressions for the degree of coherence $\mu(\rho)$ and polarization $D_p(\rho)$ defined in (2.30) and (2.31), respectively are given according to [34] by

$$\mu(\rho) = \frac{|R_x(\rho)|}{\sqrt{P_c(\rho)P_x(\rho)}}, \quad D_p(\rho) = \frac{\left\{[P_c(\rho) - P_x(\rho)]^2 + 4|R_x(\rho)|^2\right\}^{1/2}}{(P_c(\rho) + P_x(\rho))}, \quad \textit{where } 0 \le \mu(\rho) \le D_p(\rho) \le 1 \tag{3.69}$$

and for coherent (deterministic) scatterers $\mu = D_p = 1$, whereas for completely depolarized scatterers $\mu = D_p = 0$.

The covariance matrix possesses additional valuable properties for the reciprocal back-scattering case which can be demonstrated by transforming $[C_{3L}(H,V)]$ into its orthogonal representation for $\rho_{\perp} = \left(-1/\rho^*\right)$ so that

$$\left[C_{3L}\left(\rho_{\perp} = -1/\rho^*\right)\right] = \begin{bmatrix} P_c^{\perp}(\rho) & \frac{-\rho}{\rho^*}\sqrt{2}\,R_x^{\perp}(\rho) & \frac{\rho^2}{\rho^{*2}}R_c(\rho) \\ \frac{-\rho^*}{\rho}\sqrt{2}\,R_x^{\perp}(\rho)^* & 2P_x(\rho) & \frac{-\rho}{\rho^*}\sqrt{2}\,R_x(\rho)^* \\ \frac{\rho^{*2}}{\rho^2}R_c(\rho)^* & \frac{-\rho^*}{\rho}\sqrt{2}\,R_x(\rho) & P_c(\rho) \end{bmatrix} \tag{3.70}$$

leading to the following inter-channel relations

$$P_c\left(\rho_{\perp} = -1/\rho^*\right) = P_c^{\perp}(\rho) \qquad \left|R_x\left(\rho_{\perp} = -1/\rho^*\right)\right| = |R_x^{\perp}(\rho)| \tag{3.71}$$

and the symmetry relations

$$P_x\left(\rho_{\perp} = -1/\rho^*\right) = P_x(\rho) \qquad \left|R_c\left(\rho_{\perp} = -1/\rho^*\right)\right| = |R_c(\rho)| \tag{3.72}$$

Similar, but not identical relations, could be established for the Pauli-Coherency Matrix $[C_{3P}(\rho)]$, which are not presented here. There exists another polarimetric covariance matrix representation in terms of the so-called polarimetric inter-correlation parameters σ_0, ρ, δ, β, γ, and ε, where according to [19, Chapter 5]

$$[C_{3L}] = \begin{bmatrix} 1 & \beta\sqrt{\delta} & \rho\sqrt{\gamma} \\ \beta^*\sqrt{\delta} & \delta & \varepsilon\sqrt{\gamma\,\delta} \\ \rho^*\sqrt{\gamma} & \varepsilon^*\sqrt{\gamma\,\delta} & \gamma \end{bmatrix} \tag{3.73}$$

with the polarimetric inter-correlation parameters σ_0, ρ, δ, β, γ, and ε defined as

$$\begin{aligned} &\sigma_0 = \left\langle |S_{HH}|^2 \right\rangle && \rho = \frac{\left\langle S_{HH} S_{VV}^* \right\rangle}{\sigma_0\sqrt{\gamma}} = \frac{\left\langle S_{HH} S_{VV}^* \right\rangle}{\sqrt{\left\langle |S_{HH}|^2 \right\rangle \left\langle |S_{VV}|^2 \right\rangle}} \\ &\delta = 2\frac{\left\langle |S_{HV}|^2 \right\rangle}{\sigma_0} && \beta = \frac{\sqrt{2}\left\langle S_{HH} S_{HV}^* \right\rangle}{\sigma_0\sqrt{\delta}} = \frac{\left\langle S_{HH} S_{HV}^* \right\rangle}{\sqrt{\left\langle |S_{HH}|^2 \right\rangle \left\langle |S_{HV}|^2 \right\rangle}} \\ &\gamma = 2\frac{\left\langle |S_{VV}|^2 \right\rangle}{\sigma_0} && \varepsilon = \frac{\sqrt{2}\left\langle S_{HV} S_{VV}^* \right\rangle}{\sigma_0\sqrt{\delta\gamma}} = \frac{\left\langle S_{HV} S_{VV}^* \right\rangle}{\sqrt{\left\langle |S_{HV}|^2 \right\rangle \left\langle |S_{VV}|^2 \right\rangle}} \end{aligned} \tag{3.74}$$

This completes the introduction of the pertinent polarimetric matrix presentations, commonly used in radar polarimetry and in polarimetric SAR interferometry, where in addition the polarimetric interference matrices need to be introduced as shown in [19], after introducing briefly basic concepts of radar interferometry in [70].

3.12 Alternate Matrix Representations

In congruence with the alternate formulations of the of the polarization properties of electromagnetic waves, there also exist the associated alternate tensorial (matrix) formulations related to the *'four-vector Hamiltonian'* and *'spinorial'* representations as pursued by Zhivotovsky [109], and more recently by Bebbington [32]. These formulations representing most essential tools for dealing with the *'general bi-static (non-symmetric) scattering cases'* for both the coherent *(3-D Poincaré and 3-D Polarization spheroid)* and partially coherent *(4-D Zhivotovsky sphere and spheroid)* interactions, are not further pursued here; but these 'more *generalized treatments'* of radar polarimetry deserve our fullest attention.

4. Polarimetric Radar Optimization for the Coherent Case

The optimization of the scattering matrices, derived for the mono-static case is separated into two distinct classes. The first one, dealing with the optimization of $[S]$, $[G]$, and $[K]$, for the coherent case results in the formulation of '*Kennaugh's target matrix characteristic operator and tensorial polarization fork*' and the associated renamed '*Huynen Polarization Fork*' concept plus the '*co/cross-polarization power density plots*' and the 'co/cross-polarization phase correlation plots', also known as the van Zyl [79, 71] and the Agrawal plots[78, 90], respectively, in the open literature. The second one, presented in Chapter 5, deals with the optimization for the partially polarized case in terms of the '*lexicographic and the Pauli-based covariance matrices,* $[C_L]$ *and* $[C_P]$, *respectively*', as introduced in Sections 3.7 to 3.10, resulting in the '*Cloude target decomposition theorems*' and the Cloude-Pottier [27, 57, 58] supervised and unsupervised '*Polarimetric Entropy* H, *Anisotropy* A, *and* $\overline{\alpha}$*-Angle Descriptors*'. In addition, the '*polarimetric contrast optimization procedure*' dealing with the separation of the desired polarimetric radar target versus the undesired radar clutter returns of which the alternate lexicographic and Pauli-based covariance matrix optimization procedures deserve special attention next to the coherent $[S]$ and partially coherent $[K]$ matrix cases.

4.1 Formulation of the Mono-Static Radar Optimization Procedure according to Kennaugh for the Coherent Case

Kennaugh was the first to treat the mono-static polarimetric radar optimization procedure (see Fig. 4.1) for optimizing (3.9) according to the BSA formulation

$$\mathbf{E}^s(\mathbf{r}) = [S]\,\mathbf{E}^{i*}(\mathbf{r}) \tag{4.1}$$

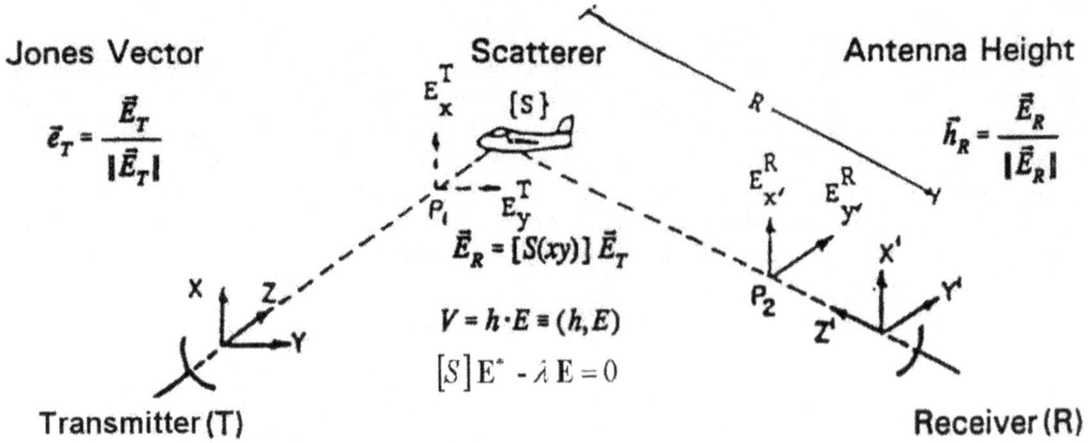

Fig. 4.1 BSA Optimization According to Kennaugh

Boerner, W.-M. (2007) Basics of SAR Polarimetry II. In *Radar Polarimetry and Interferometry* (pp. 4-1 – 4-30). Educational Notes RTO-EN-SET-081bis, Paper 4. Neuilly-sur-Seine, France: RTO. Available from: http://www.rto.nato.int/abstracts.asp.

but with the received field $\mathbf{E}^r(\mathbf{r})$ being so aligned with the incident field $\mathbf{E}^i(\mathbf{r})$ with the reversal of the scattered versus incident coordinates of the BSA system resulting in Kennaugh's psuedo-eigenvalue' [4] problem of

$$Opt\{[S]\} \text{ such that } [S]\mathbf{E}^* - \lambda\,\mathbf{E} = 0 \tag{4.2}$$

The rigorous solution to this set of 'con-similarity eigenvalue' problems was unknown to the polarimetric radar community until the late 1980's, when Lüneburg [54], rediscovering the mathematical tools [116, 117], derived a rigorous but mathematically rather involved method of the associated con-similarity eigenvalue problem, not further discussed here, but we refer to Lüneburg's complete treatment of the subject matter in [52, 53]. Instead, here Chan's [77] *'Three-Step Solution'*, as derived from Kennaugh's original work [4], is adopted.

Three Step Procedure according to Chan [77]

By defining the polarimetric radar brightness (polarization efficiency, polarization match factor) formation according to (3.26) and (3.27) retaining the factor $1/2$ (not contained in the 1983 IEEE Standard and in Mott's textbook) [76, 102] as

$$P_R = |V_R|^2 = \frac{1}{2}\left|\mathbf{h}^{r^T}\,[\mathbf{S}]\,\mathbf{E}^i\right|^2 \tag{4.3}$$

in terms of the terminal voltage V_R, being expressed in terms of the normalized transceiver antenna height $\mathbf{h}^r$ and the incident field $\mathbf{E}^i$, as defined in Mott [76] and in [19], by

$$V_R = \mathbf{h}^{r^T}\,\mathbf{E}^s = \mathbf{h}^{r^T}[S]\mathbf{E}^i \quad \textbf{with} \quad \mathbf{h}^r = \frac{\mathbf{E}^r}{\|\mathbf{E}^r\|} \tag{4.4}$$

so that the total energy density of the scattered wave $\mathbf{E}^s$, may be defined by

$$W = \mathbf{E}^{s^+}\mathbf{E}^s = ([S]\mathbf{E}^i)^+([S]\mathbf{E}^i) = \mathbf{E}^{i^+}([S]^+[S])\mathbf{E}^i = \mathbf{E}^{i^+}[G]\mathbf{E}^i \tag{4.5}$$

where $[G] = [S]^\dagger[S]$ defines the Graves power density matrix [7], first introduced by Kennaugh [4, 5].

Step 1

Because the solution to the *'pseudo-eigenvalue problem'* of (4.2) was unknown at that time (1954 until 1984); and, since $[S]$ could be, in general, non-symmetric and non-hermitian, Kennaugh embarked instead in determining the *'optimal polarization states'* from optimizing the power density matrix so that

$$[G]\mathbf{E}'_{\mathbf{opt}} - \nu\,\mathbf{E}'_{\mathbf{opt}} = 0 \tag{4.6}$$

for which real positive eigenvalues $\nu_1 = |\lambda_1|^2$ and $\nu_2 = |\lambda_2|^2$ exist for all matrices $[S]$ since $[G]$ is Hermitian positive semidefinite so that

$$\nu_{1,2} = \frac{1}{2}\left\{Trace[G] \pm \left[Trace^2[G] - 4Det[G]\right]^{\frac{1}{2}}\right\} \tag{4.7}$$

where

$$\nu_1 + \nu_2 = \textbf{invariant} = Trace[G] = Span[S] = [S_{HH}]^2 + [S_{HV}]^2 + [S_{VH}]^2 + [S_{VV}]^2 = \kappa_4 \tag{4.8}$$

$$\nu_1 \cdot \nu_2 = \textbf{invariant} = Det[G] = (Det[S])(Det[S])^* = (S_{HH}S_{VV} - S_{HV}S_{VH})(S_{HH}S_{VV} - S_{HV}S_{VH})^* \tag{4.9}$$

For the mono-static, reciprocal symmetric $[S]$, above equations reduce with $S_{HV} = S_{VH}$ to

$$\nu_1 + \nu_2 = \textbf{invariant} = Trace[G] = Span[S] = [S_{HH}]^2 + 2[S_{HV}]^2 + [S_{VV}]^2 = \kappa_3 \tag{4.10}$$

and

$$\nu_1 \cdot \nu_2 = \textbf{invariant} = Det[G] = (Det[S])(Det[S])^* = \left(S_{HH}S_{VV} - |S_{HV}|^2\right)\left(S_{HH}S_{VV} - |S_{HV}|^2\right)^* \tag{4.11}$$

In order to establish the connection between the coneigenvalues of equation (4.2) and the eigenvalues of $[G]$ in (4.6), one may proceed to take the complex conjugate of (4.2) and insert back in (4.2). Equation (4.2) has orthogonal solutions if and only if $[S]$ is symmetric. The inverse step is much more difficult to prove and needs among others the symmetry of $[S]$, which provides another topic for future research.

As a result of these relations, Kennaugh defined the *'effective polarimetric radar cross-section'* ε_{K_4}, also known as '***Kennaugh's Polarimetric Excess*** ε_{K_4} ', in [118], where

$$\varepsilon_{K_4} = Span[S] + 2|Det[S]| \tag{4.12}$$

which comes automatically into play (also in the present formulation) when representations on the Poincare sphere are considered, which reduces to ε_{K_3} for the mono-static reciprocal case. It plays an essential role in Czyz's alternate formulation of the *'theory of radar polarimetry'* [110], derived from a spinorial transformation concept on the *'generalized polarization sphere'*, being studied in more depth by Bebbington [32].

Step 2

Using the resulting solutions for $\nu_{1,2}$ for the known $[G]=[S]^{+}[S]$ and $[S]$, the optimal transmit polarization states $\mathbf{E}^{t}_{\mathbf{opt}1,2}$ and optimal scattered waves $\mathbf{E}^{s}_{\mathbf{opt}1,2}$ can be determined as

$$\mathbf{E}^{s}_{\mathbf{opt}1,2}=[S]\,\mathbf{E}^{t}_{\mathbf{opt}1,2} \tag{4.13}$$

Step 3

The received optimal antenna height $\mathbf{h}^{r}_{\mathbf{opt}}$ is then derived from (4.4) as

$$\mathbf{h}^{r}_{\mathbf{opt}}=\frac{\mathbf{E}^{s^*}_{\mathbf{opt}}}{\left\|\mathbf{E}^{s}_{\mathbf{opt}}\right\|}=\frac{\left([S]\mathbf{E}^{t}_{\mathbf{opt}}\right)^*}{\left\|[S]\mathbf{E}^{t}_{\mathbf{opt}}\right\|} \tag{4.14}$$

which defines the *'polarization match'* for obtaining maximum power in terms of the polarimetric brightness function (4.4) introduced by Kennaugh in order to solve the polarimetric radar problem [4].

There exist several alternate methods of determining the optimal polarization states either by implementing the *'generalized complex polarization ρ'* transformation, first pursued by Boerner et al. [13]; the *'con-similarity transformation method'* of Lüneburg [52, 53] , the *'spinorial polarization sphere transformations'* of Bebbington [32], and more recently the *'Abelian group method'* of Yang [104 - 105]. It would be worthwhile to scrutinize the various approaches, which should be a topic for future research.

4.2 The Generalized ρ - Transformation for the Determination of the Optimal Polarization States by Boerner using the Critical Point Method

Kennaugh further pioneered the *'polarimetric radar optimization procedures'* by transforming the optimization results on to the polarization sphere, and by introducing the co-polarized versus cross-polarized channel decomposition approach [4] which were implemented but not further pursued by Huynen [9]. Boerner et al. [31, 82], instead, proposed to implement the complex polarization ratio ρ transformation in order to determine the pairs of maximum/minimum back-scattered powers in the co/cross-polarization channels and optimal polarization phase instabilities (cross-polar saddle extrema) by using the *'critical point method'* pioneered in [82]. Assuming that the scattering matrix $[S(HV)]$ is transformed to any other ortho-normal basis {A B} such that

$$[S'(AB)]=\begin{bmatrix} S'_{AA} & S'_{AB} \\ S'_{BA} & S'_{BB}\end{bmatrix}=[U]^T\begin{bmatrix} S_{HH} & S_{HV} \\ S_{VH} & S_{VV}\end{bmatrix}[U] \tag{4.15}$$

with $[U]$ given by (2.23); and $S_{HV}=S_{VH}$, $S'_{AB}=S'_{BA}$ for the mono-static case, the polarimetric radar brightness equation becomes

$$P=\frac{1}{2}\,|V|^2=\frac{1}{2}\left|\mathbf{E}^{r^T}[S]\,\mathbf{h}^{t}\right|^2=\frac{1}{2}\left|\mathbf{E}^{r'^T}\left[S'\right]\mathbf{h}^{t'}\right|^2 \tag{4.16}$$

where the prime ' refers to any new basis {A B} according to (4.15).

By implementation of the Takagi theorem [116], the scattering matrix $[S'(AB)]$ can be diagonalized [52] so that

$$[S'(AB)]=\begin{bmatrix} S'_{AA} & 0 \\ 0 & S'_{BB}\end{bmatrix}=\begin{bmatrix} \lambda_1 & 0 \\ 0 & \lambda_2\end{bmatrix}=[S_d] \tag{4.17}$$

$$\lambda_1 = S'_{AA}(\rho_1) = \left(1+\rho_1\rho_1^*\right)^{-1}\left(S_{HH} + 2\rho_1 S_{HV} + \rho_1^2 S_{VV}\right)\exp(2j\psi_1) = |\lambda_1| \exp(2j\phi_1) \quad (4.18)$$

$$\lambda_2 = S'_{BB}(\rho_1) = \left(1+\rho_1\rho_1^*\right)^{-1}\left(\rho_1^{*2} S_{HH} - 2\rho_1^* S_{HV} + S_{VV}\right)\exp(2j\psi_4) = |\lambda_2| \exp(2j\phi_2) \quad (4.19)$$

as shown in [31].

Determination of the Kennaugh target matrix characteristic polarization states: The expression for the power returned to the co-pol and cross-pol channels of the receiver are determined from the bilinear form to become:

(i) Cross-pol Channel Minima or Nulls (n), Maxima (m) and Saddle-Optima (s):
For the Cross-pol channel power P_x *with* $\mathbf{E}^r = \mathbf{E}^{t\perp}$, expressed in terms of the antenna length $\mathbf{h}$

$$P_x = |V_x|^2 = \frac{1}{2}\left|\mathbf{h}_\perp^{'T}\,[S]\,\mathbf{h}'\right|^2 = \frac{1}{\left(1+\rho'\rho'^*\right)^2}\left(|\lambda_1|^2\rho'\rho'^* - \lambda_1\lambda_2^*\rho'^{*2} - \lambda_1^*\lambda_2\rho'^2 + |\lambda_2|^2\rho'\rho'^*\right)$$

(4.20)

so that for the cross-pol nulls ($\rho'_{xn1,2}$), for the cross-pol maxima ($\rho'_{xm1,2}$), and for the cross-pol saddle optima ($\rho'_{xs1,2}$), according to the critical point method introduced in [82],

$$\rho'_{xn1,2} = 0,\infty \qquad \rho'_{xm1,2} = \pm j\left(\frac{\lambda_1\lambda_2^*}{\lambda_1^*\lambda_2}\right)^{\frac{1}{4}} = \pm\exp(j(2\nu+\pi/2)) \qquad \rho'_{xs1,2} = \pm\left(\frac{\lambda_1\lambda_2^*}{\lambda_1^*\lambda_2}\right)^{\frac{1}{4}} = \pm\exp(j(2\nu))$$

(4.21)

where

$$\rho'_{xn1}\rho'^*_{xn2} = -1 \qquad \rho'_{xm1}\rho'^*_{xm2} = -1 \qquad \rho'_{xs1}\rho'^*_{xs2} = -1 \qquad (4.22)$$

which states that there exist three pairs of orthogonal polarization states, the cross-pol minima ($\rho'_{xn1,2}$), the cross-pol maxima ($\rho'_{xm1,2}$), and the cross-pol saddle optima ($\rho'_{xs1,2}$), which are located pair-wise at antipodal points on the polarization sphere so that the lines joining the orthogonal polarization states are at right angles to each other on the polarization sphere[82].

(ii)Co-pol Channel Maxima ($\rho'_{cm1,2}$) and Minima or Nulls ($\rho'_{cn1,2}$):
For the function of the power P_c *with* $\mathbf{E}^r = \mathbf{E}^t$, return to the co-pol channel (c)

$$P_c = \frac{1}{2}|V_c|^2 = \frac{1}{2}\left|\mathbf{h}^{'T}\,[S_d]\,\mathbf{h}'\right|^2 = \frac{1}{\left(1+\rho'\rho'^*\right)^2}\left(|\lambda_1|^2 + \lambda_1\lambda_2^*\rho'^{*2} + \lambda_1^*\lambda_2\rho'^2 + |\lambda_2|^2\rho'^2\rho'^{*2}\right)$$

(4.23)

the critical points are determined from

$$\rho'_{cm1} = \rho'_{xn1} = 0 \qquad \rho'_{cm2} = \rho'_{xn2} = \infty, \quad where \quad \rho'_{cm1}\rho'_{cm2} = -1 \qquad (4.24)$$

and the co-pol maxima are identical to the cross-pol nulls as was first established by Kennaugh [4, 5], and utilized by Huynen [9]. In addition the critical points for the co-pol-null minima or nulls ($\rho'_{cn1,2}$) are determined from (4.23) to be

$$\rho'_{cn1,2} = \pm\left(\frac{\lambda_1}{\lambda_2}\right) = \pm\left(\frac{|\lambda_1|}{|\lambda_2|}\right)^{\frac{1}{2}} \exp(j(2\nu + \pi/2)) \tag{4.25}$$

and it can be shown from above derivations that the co-pol-null minima ρ'_{cn1} and ρ'_{cn2} lie in a plane spanned by the co-pol-maxima (cross-pol-minima) and the cross-pol-maxima and the angle between the origin of the polarization sphere and the two co-pol-nulls is bisected by the line joining the orthogonal pair of co-pol-maxima (cross-pol-minima) defining the target matrix critical angle 2x2γ as shown first by Kennaugh [4] leading to his tensorial polarization fork formulation.

(iii)<u>Orthogonality Conditions with Corresponding Power Returns</u>:

The three pairs of cross-pol-extrema, the cross-pol nulls ($\rho'_{xn1,2}$) being identical to co-pol maxima ($\rho'_{cm1,2}$), the cross-pol maxima ($\rho'_{xm1,2}$) and cross-pol saddle optima ($\rho'_{xs1,2}$), satisfy the orthogonality conditions of (4.22) and (4.24) which implies that they are located each at anti-podal locations on the polarization sphere. We note that the co-pol maxima' consist of one absolute maximum and an orthogonal local maximum. The corresponding co/cross-polar power returns become

$$Min\{P_x\} = P_{xn1}(\rho'_{xn1}) = P_{xn2}(\rho'_{xn2}) = 0 \;;$$

$$Max\{P_c\}:\; P_{cm1}(\rho'_{cm1}) = |\lambda_1|^2 \quad P_{cm2}(\rho'_{cm2}) = |\lambda_2|^2 \;;$$

$$Max\{P_x\}:\; P_x(\rho'_{xm1,2}) = \frac{1}{4}(|\lambda_1|^2 + |\lambda_2|^2); \tag{4.26}$$

$$Min\{P_c\}:\; P_c(\rho'_{cn1,2}) = 0;$$

$$Sad\{P_x\}:\; P_x(\rho'_{xs1,2}) = \frac{1}{4}(|\lambda_1|^2 - |\lambda_2|^2)$$

The resulting co/cross-polar extrema are plotted on the polarization sphere shown in Fig. 4.2

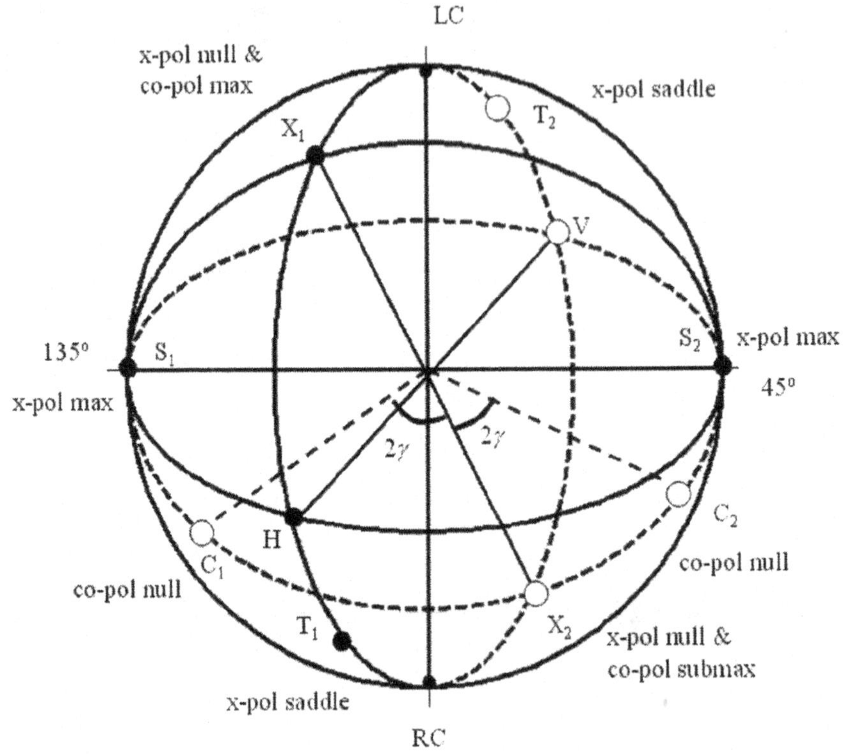

Fig. 4.2 Co/cross-polar Extrema

4.3 The Kennaugh Target Characteristic Scattering Matrix Operator, and the Polarization Fork according to Boerner

Kennaugh was the first to recognize that the orthogonal pairs (X_1, X_2) of the cross-pol nulls ($\rho'_{xn1,2}$) or co-pol maxima ($\rho_{cm1} = \rho_{xn1}, \rho_{cm2} = \rho_{xn2}$) and the pair ($S_1$, S_2) of cross-pol maxima ($\rho'_{xm1,2}$) lie in one main cross-sectional plane of the polarization sphere also containing the pair (C_1, C_2) of non-orthogonal co-pol nulls ($\rho_{cn1,2}$), where the angle 4γ between the two co-pol null vectors on the Poincaré sphere is bisected by the line joining the two co-pol maxima (cross-pol nulls). These properties were first recognized explicitly and utilized by Kennaugh for defining his "Spinorial Polarization Fork", used later on by Huynen to deduce his *'Polarization Fork'* concept.

However, Boerner et al. [13, 25, 81, 31, 82], by implementing the complex polarization ratio transformation, were able to relate the polarization state coordinates $P(\rho')$ on the Polarization sphere directly to the corresponding ρ' on the complex polarization ratio plane. Then according to [82], each point ρ' of the complex plane can be connected to the *'zenith (LC)'* of the polarization sphere, resting tangent to the complex plane in its *'origin 0'* of the *'nadir (RC)'*, by a straight line that intersects the sphere at one arbitrary point , where the *'nadir (RC)'* corresponds to the *'origin 0'* of the ρ' -plane, the *'zenith (z)'* to the ρ' -circle at *'infinity (∞'* and the equator representing linear polarization states. Any two orthogonal polarization states are antipodal on the sphere, like *'zenith (left-circular)'*, and *'nadir (right-circular)'*. Utilizing this property, Boerner and Xi [31] were able to associate uniquely three pairs of orthogonal polarization states at right angle (bi-orthogonal) on the polarization sphere; i.e. the anti-podal points S_1, S_2 ($\rho'_{xm1,2}$) and T_1, T_2 ($\rho'_{xs1,2}$), with $\overline{S_1S_2}$ and $\overline{T_1T_2}$ being perpendicular to one another (bi-orthogonal); and similarly to the line $\overline{X_1X_2}$ joining X_1 (nadir: $\rho'_{xn1} = \rho'_{cm1} = 0$) and X_2 (zenith: $\rho'_{xn2} = \rho'_{cm2} = \infty$); where the co-pol nulls ($\rho'_{cn1,2}$) lie on the same main circle on the complex plane of $\rho'_{cn1,2}$ and $\rho'_{xn1,2}$ so that their corresponding points C_1 and C_2 are symmetric about the diameter $\overline{X_1X_2}$ which bisects the angle between C_1, 0, C_2 known as the Kennaugh target matrix characteristic angle (2x2 γ). The *'great cross-sectional plane'* containing $\overline{S_1S_2} \perp \overline{X_1X_2}$ and C_1, C_2 is denoted as the *'Kennaugh target matrix characteristic plane'* with corresponding great circle being the *'Kennaugh target matrix characteristic circle'*. Fig. 4.3 define the representations of the Poincaré sphere for the general and standardized polarization fork (Huynen), respectively, including the proper definitions of *'Huynen's Geometrical Parameters (φ-target orientation angle; γ-target ship angle; τ-target ellipticity angle,* $\rho = \tan\alpha \exp(j\delta)$ *'*, next to *'Kennaugh's target matrix characteristic angle* γ *'*.

This concludes the description of the *'Kennaugh polarimetric target matrix characteristic operator'*; which was coined *'the polarization fork'* by Huynen [9].

4.4 Huynen's Target Characteristic Operator and the Huynen Polarization Fork

Huynen [9], utilizing Kennaugh's prior studies [4, 5], elaborated on polarimetric radar phenomenologies extensively, and his *"Dissertation of 1970: Phenomenological Theory of Radar Targets"* [9], re-sparked international research on *Radar Polarimetry*, commencing with the studies by Poelman [10, 11], Russian studies by Kanareykin [122], Potekhin [123], and others [1].

Huynen cleverly reformulated the definition of the polarization vector, as stated in [9], so that group-theoretic Pauli-spin matrix concepts may favorably be applied which also serve for demonstrating the orientation angle invariance which Huynen coined *'de-psi-ing (de-ψ-ing)'* using ψ for denoting the relative polarization ellipse orientation angle. Here, we prefer to divert from our notation by rewriting the parametric definition of the polarization vector

$$\mathbf{p}(|E|,\phi,\psi,\chi) = |E|\exp(j\phi)\begin{bmatrix}\cos\psi & -\sin\psi\\ \sin\psi & \cos\psi\end{bmatrix}\begin{bmatrix}\cos\chi\\ -j\sin\chi\end{bmatrix}$$

as

$$\mathbf{p}(a,\alpha,\phi,\tau) = a\exp(j\alpha)\begin{bmatrix}\cos\phi & -\sin\phi\\ \sin\phi & \cos\phi\end{bmatrix}\begin{bmatrix}\cos\tau\\ -j\sin\tau\end{bmatrix} \tag{4.27}$$

which with the use of the Pauli-spin matrices $[\sigma_i]$ defined in (2.14), the Huynen quaternion group definitions may be re-expressed as $[I]=[\sigma_0]$, $[J]=-j[\sigma_3]$, $[K]=j[\sigma_2]$, $[L]=-j[\sigma_1]$

$$\mathbf{p}(a,\alpha,\phi,\tau) = a\exp(j\alpha)\exp(\phi[J])\exp(\tau[K])\begin{bmatrix}1\\0\end{bmatrix} \tag{4.28}$$

with $\exp(\phi[J]) = \cos\phi[I] + \sin\phi[J]$ and $\exp(\tau[K]) = \cos\tau[I] + \sin\tau[K]$

In this notation the orthogonal polarization vector $\mathbf{p}_\perp$ becomes

$$\mathbf{p}_\perp(a_\perp,\alpha_\perp,\phi,\tau) = a_\perp\exp(j\alpha_\perp)\exp\{(\phi+\frac{\pi}{2})[J]\}\exp(-\tau[K])\begin{bmatrix}1\\0\end{bmatrix} \tag{4.29}$$

so that $\mathbf{p}\cdot\mathbf{p}^* = a^2, \quad \mathbf{p}\cdot\mathbf{p}_\perp^* = 0$.

Utilizing this notation, the transformed matrix $[S'(AB)]$ becomes

$$[S'(AB)] = [U]^T [S(HV)][U] \tag{4.30}$$

with the orthonormal transformation matrix $[U]$ defined in (2.23), which may be recasted with $\mathbf{m} = \mathbf{p}_m$ the so-called maximum or null polarization as defined in [31], into

$$[U] = [\mathbf{m}\ \mathbf{m}_\perp] \tag{4.31}$$

Because of the orthonormal properties of $\mathbf{m}$ and $\mathbf{m}_\perp$, which satisfy the con-similarity eigenvalue equation [82], the off-diagonal elements of $[S'(AB)]$ vanish. This in turn can be used to solve for ρ in (2.23), and hence for $\mathbf{m}$ and $\mathbf{m}_\perp$, without solving the consimilarity eigenvalue problem of (4.6). The complex eigenvalues $\rho_{xm1,2}$, defined in (4.26) are renamed as $s_{1,2}$ and were defined by Huynen as

$$s_1 = m\exp\{2j(\upsilon+\beta)\} \qquad s_2 = m\tan^2\gamma\exp\{-2j(\upsilon-\beta)\} \tag{4.32}$$

so that $[S'(AB)]$ of (4.30) becomes

$$[S'(AB)] = [U^*(\mathbf{m},\mathbf{m}_\perp)]\begin{bmatrix} m\exp\{2j(\upsilon+\beta)\} & 0 \\ 0 & m\tan^2\gamma\exp\{-2j(\upsilon-\beta)\}\end{bmatrix}[U^*(\mathbf{m},\mathbf{m}_\perp)]^T \tag{4.33}$$

where $m = \sigma_K, \gamma, \psi, \tau_m, \upsilon,$ *and* β are the Huynen parameters, and m denoting the target matrix magnitude, may be identified to be *"Kennaugh's polarimetric excess* σ_K *"* defined in (4.12); and $\mathbf{m}(\psi,\tau_m)$ may be re-normalized as

$$\mathbf{m}(\psi,\tau_m) = \begin{bmatrix}\cos\psi & -\sin\psi \\ \sin\psi & \cos\psi\end{bmatrix}\begin{bmatrix}\cos\tau_m \\ -j\sin\tau_m\end{bmatrix} = \exp(\psi[\mathbf{J}])\exp(\tau_m[\mathbf{K}])\begin{bmatrix}1\\0\end{bmatrix} \tag{4.34}$$

and finally with

$$\begin{bmatrix} m\exp\{2j(\upsilon+\beta)\} & 0 \\ 0 & m\tan^2\gamma\exp\{-2j(\upsilon-\beta)\}\end{bmatrix} = m\exp(j\beta)\begin{bmatrix}1 & 0\\ 0 & \tan^2\gamma\end{bmatrix}\exp(\upsilon[\mathbf{L}]) \tag{4.35}$$

and $[L]$ representing the third modified Pauli-spin matrix, satisfying in Huynen's notation

$$[L] = [J][K] = -[K][J] - \begin{bmatrix}-j & 0\\ 0 & j\end{bmatrix}, \qquad [L]^2 = -[I] \tag{4.36}$$

we obtain Huynen's target matrix characteristic operator

$$[\mathbf{H}(m,\gamma,\beta;\phi_m,\upsilon,\tau_m)] = [U^*(\psi,\tau_m,\upsilon)]\, m\begin{bmatrix}1 & 0\\ 0 & \tan^2\gamma\end{bmatrix}[U^*(\psi,\tau_m,\upsilon)]^T\exp(j\beta) \tag{4.37}$$

where

$$[U(\psi,\tau_m,\upsilon)]\exp(\psi[\mathbf{J}])\exp(\tau_m[\mathbf{K}])\exp(\upsilon[\mathbf{L}]) \qquad (4.38)$$

representing the Eulerian rotations with 2ψ, $2\tau_m$, 2υ about the bi-orthogonal polarization axes $\overline{S_1S_2}$ (connecting the two cross-pol maxima), $\overline{X_1X_2}$ (connecting the two cross-pol nulls, or equivalently, co-pol maxima), and $\overline{T_1T_2}$ (connecting the two saddle optima), respectively, with more detail given in Boerner and Xi [31]. Huynen pointed out the significance of the relative target matrix orientation angle $\Phi = \phi - \psi$, where ϕ denotes the antenna orientation angle, and that from definition of $\mathbf{m}(\psi,\tau_m)$ in (4.34), it can be shown that it can be eliminated from the scattering matrix parameters and incorporated into the antenna polarization vectors (*'de-psi-ing: de-ψ-ing'*) , and that the Huynen parameters are orientation independent, which was more recently analyzed in depth by Pottier [58]. The Eulerian angle are indicators of a scattering matrix's characteristic structure with υ denoting the so-called *'skip-angle'* related to multi-bounce scattering (single versus double), τ_m denotes the helicity-angle and is an indicator of target symmetry $\tau_m = 0$ or non-symmetry, and β is the absolute phase which is of particular relevance in polarimetric radar interferometry.

$$[S] = [U^*(\rho)]\exp(\nu[L]^*)\, m \begin{bmatrix} 1 & 0 \\ 0 & \tan^2\gamma \end{bmatrix} \exp(\nu[L]^*)^T [U^*(\rho)]^T \exp(j\xi)$$

$$[U(\rho)]^* = \frac{1}{\sqrt{1+\rho\rho^*}} \begin{bmatrix} e^{-j\psi_1} & -\rho(\phi_m,\tau_m)e^{-j\psi_4} \\ \rho^*(\phi_m,\tau_m)e^{-j\psi_1} & e^{-j\psi_4} \end{bmatrix}$$

$$m = |\lambda_1|$$

$$\rho'_{xn1} = H = \rho'_{cn1}, \quad \rho'_{xn2} = V = \rho'_{cn2}$$

$$\rho'_{cn1,2} = \pm\tan(\pi/2 - \gamma)\exp[j(2\gamma + \pi/2)]$$

$$\rho'_{xm1,2} = \pm\exp[j(2\gamma + \pi/2)] \quad (LR)$$

$$\rho'_{xs1,2} = \pm\exp(j2\gamma) \quad (45^\circ/135^\circ)$$

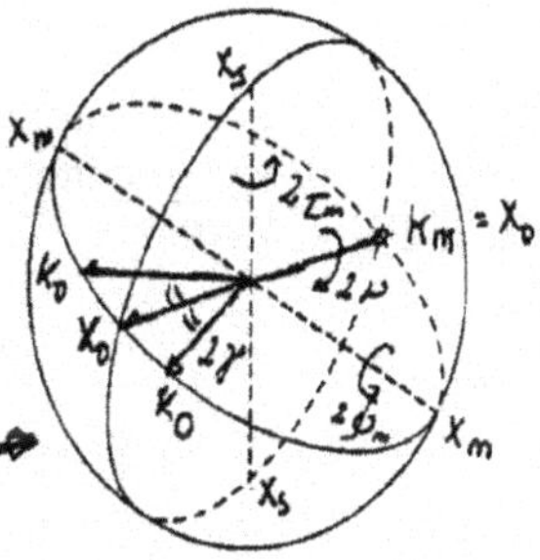

Huynen's Solution of the Polarization Fork

$$[H] = [U^*(\psi,\tau_m,\nu)]\, m \begin{bmatrix} 1 & 0 \\ 0 & \tan^2\gamma \end{bmatrix} [U^{*\prime}(\psi,\tau_m,\nu)]\exp(j\xi)$$

$$[U(\psi,\tau_m,\nu)] = e^{\psi[J]} e^{\tau_m[K]} e^{\nu[L]}$$

Fig. 4.3 Huynen's Polarization Fork (Xi-Boerner Solution)

4.5 Alternate Coherent Scattering Matrix Decompositions by Kroggager and by Cameron

Another class of scattering matrix decomposition theorems [124, 30] were recently introduced, and are also expressed in terms of the Pauli spin matrix sets $\psi_p([\sigma_i],\ i = 0,1,2,3)$, by associating elementary scattering mechanisms with each of the $[\sigma_i]$, so that for the general non-symmetric case

$$[S(a,b,c,d)] = \begin{bmatrix} a+b & c-jd \\ c+jd & a-b \end{bmatrix} = a[\sigma_0] + b[\sigma_1] + c[\sigma_2] + d[\sigma_3] \tag{4.39}$$

where a, b, c and d are all complex. Above '*coherent*' decomposition may be interpreted in terms of **four** '*elementary deterministic point target*' scattering mechanisms, viewed under a change of wave polarization basis, where

a - corresponds to single scattering from a sphere or plane surface
b - corresponds to di-plane scattering (4.40)
c - corresponds to di-plane scattering with a relative orientation of 45°
d - corresponds to anti-symmetric 'helix-type' scattering mechanisms that transform the incident wave into its orthogonal circular polarization state (helix related)

Krogager [30] and Cameron [124, 125], among others, in essence made use of this decomposition for the symmetric scattering matrix case by selecting the desirable combinations of the $[\sigma_i]$ that suits their specific model cases best.

In the '*Krogager Approach*', a symmetric matrix $[S_K]$ is decomposed into three coherent components, which display the '*physical meaning*' of '*sphere*', '*diplane*', and '*helical targets*', where

$$[S_K(a,b,d)] = \alpha [S_{sph}] + \mu \exp(j\phi)[S_{di}] + \eta \exp(j\phi)[S_{hel}] \tag{4.41}$$

with additional direct associations with the Pauli matrices defined in (2.14) given by

$$[S_{sph}] = \begin{bmatrix} 1 & 0 \\ 0 & 1 \end{bmatrix} \qquad [S_{di}] = \begin{bmatrix} 1 & 0 \\ 0 & -1 \end{bmatrix} \qquad [S_{hel}] = \begin{bmatrix} 0 & -j \\ j & 0 \end{bmatrix} \tag{4.42}$$

This decomposition is applied directly to the complex $[S]$ matrix imagery, and results into a rather efficient sorting algorithm in terms of the three characteristic '*feature sorting base scatter images*'. Using color composite presentations for the three classes then allows for the associated '*unsupervised feature sorting*'. This *feature sorting method* has been applied rather successfully in the interpretation of various geo-environmental (forestry, agriculture, fisheries, natural habitats, etc.) as well as in law enforcement and military applications.

In the '*Cameron Approach*', the matrix $[S_C(a,b,c,d)]$ is decomposed, by separating the non-reciprocal $[S_{C_{nr}}]$ from the reciprocal component $[S_{C_{rec}}] = [S_{C_{sym}}]$ via an orientation angle ϕ', and by further decomposing the latter into two further components, $[S_{C_{sym}}^{\max}]$ and $[S_{C_{sym}}^{\min}]$, with linear eigen-polarizations via the angle τ, so that

$$[S_C(a,b,c,d)] = a\cos\phi' \left\{\cos\tau \left[S_{C_{sym}}^{\max}\right] + \sin\tau \left[S_{C_{sym}}^{\min}\right]\right\} + a\sin\phi' \left[S_{C_{nr}}\right] \tag{4.43}$$

which is further analyzed in [40] and its limitations are clearly identified in [27].

Of the many other existing $[S(a,b,c,d)]$ matrix decomposition theorems, mostly derived from alternate formulations (4.39) of the Pauli spin matrix set $\psi_p([\sigma_i],\ i = 0,1,2,3)$, defined in (2.14), the three examples of the Huynen, the Krogager and the Cameron decompositions, it becomes apparent that there exists an infinum of decompositions non of which is unique and all of them are basis dependent and require *a priori* information on the scatterer scenario under investigation. Yet for specific distinct applications all of them may serve a useful purpose which highly superior to any non-polarimetric or partially polarimetric treatment.

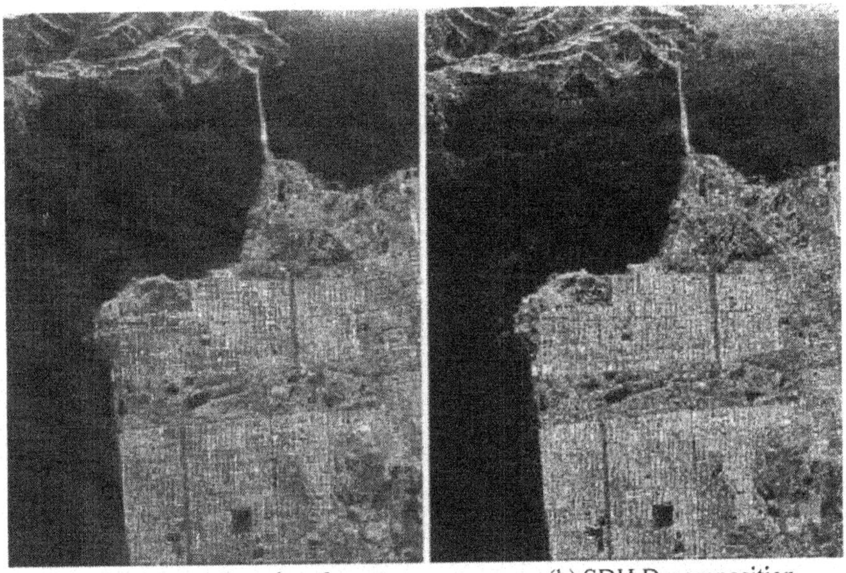

(a) San Francisco Image (b) SDH Decomposition

Fig. 4.4 Polarimetric Decompositions: Krogager, 1993. (a) Original San Francisco POLSAR image with RGB color coded by |HH-VV|, |HV| and |HH+VV|, respectively.
(b) Sphere (Blue), diplane (Red) and helix (Green) decomposition (SDH) decomposition.

4.6 Kennaugh Matrix Decomposition of Huynen's Matrix Vector Characteristic Operator

The '*Kennaugh target matrix characteristic operator*' can also be derived from the Kennaugh matrix, as was shown by Boerner et al. [78, 90, 31, 82]; using the Lagrangian multiplier method. A more elegant method was recently devised by Pottier in order to highlight the importance of Huynen's findings on the '*target orientation ψ invariance*' for both the Kennaugh and the Lexicographic Covariance matrix representations; and most recently by Yang [119 - 121], who shed more light into the properties of the '*equi-power-loci*' as well as '*equi-correlation-phase-loci*' which deserve careful future attention but will not be further analyzed here.

Instead, we return to Huynen [9], who provided further phenomenological insight into the properties of the Kennaugh matrix $[K]$, by redefining $[S]$ for the symmetric case in terms of the limited set of Pauli spin matrices.

$$[S(a,b,c)] = \begin{bmatrix} a+b & c \\ c & a-b \end{bmatrix} = a[\sigma_0] + b[\sigma_1] + c[\sigma_2] \tag{4.44}$$

so that with the formal relation of $[K]$ with $[S]$, obtained via a Kroenecker product multiplication as

$$[K] = 2[A]^{T-1}\left([S]\otimes[S^*]\right)[A]^{-1} \tag{4.45}$$

insertion of (4.44) into (4.45) yields, using Huynen's notation

$$[K_\psi] = \begin{bmatrix} (A_0+B_0) & F_\psi & C_\psi & H_\psi \\ F_\psi & (-A_0+B_0) & G_\psi & D_\psi \\ C_\psi & G_\psi & (A_0+B_1) & E_\psi \\ H_\psi & D_\psi & E_\psi & (A_0-B_0) \end{bmatrix} \tag{4.46}$$

where

$$\begin{aligned} |a|^2 &= 2A_0 & ab^* &= C - jD \\ |b|^2 &= B_0 + B_1 & bc^* &= E + jF \\ |c|^2 &= B_0 - B_1 & ac^* &= H + jG \end{aligned} \tag{4.47}$$

Recognizing that H becomes zero for proper '*de-ψ-ing*' of $[S]$ as defined in $[H]$ of (4.37) by removing the ψ rotational dependence, and inserting it into the antenna descriptors, he was able to redefine his Kennaugh matrix coefficients such that

$$\begin{aligned} H_\psi &= C\sin 2\psi & C_\psi &= C\cos 2\psi \\ G_\psi &= G\cos 2\psi - D\sin 2\psi & E_\psi &= E\cos 4\psi + F\sin 4\psi \\ D_\psi &= G\sin 2\psi + D\cos 2\psi & F_\psi &= -E\sin 4\psi + F\cos 4\psi \end{aligned} \tag{4.48}$$

so that the '*de-ψ-ed*' $[K]$ becomes, by removing the ψ-dependence from $[K]$ and incorporating it into the antenna polarization and Stokes' vectors, respectively, such that

$$[K_{de-\psi}] = \begin{bmatrix} (A_0 + B_0) & F & C & H = 0 \\ F & (-A_0 + B_0) & G & D \\ C & G & (A_0 + B_1) & E \\ H = 0 & D & E & (A_0 - B_0) \end{bmatrix} \tag{4.49}$$

and expressed in terms of the Huynen parameters

$$\begin{aligned} A_0 &= Qf\cos^2 2\tau_m & G &= Qf\sin 4\tau_m \\ B_0 &= Q(1 + \cos^2 2\gamma - f\cos^2 2\tau_m) & B_1 &= Q(1 + \cos^2 2\gamma - f(1 + \sin^2 2\tau_m)) \\ C &= 2Q\cos 2\gamma\cos 2\tau_m & D &= Q\sin^2 2\gamma\sin 4\upsilon\cos 2\tau_m \\ F &= 2Q\cos 2\gamma\sin 2\tau_m & E &= -Q\sin^2 2\gamma\sin 4\upsilon\sin 2\tau_m \\ Q &= m^2(8\cos^4\gamma)^{-1} & f &= 1 - \sin^2 2\gamma\sin^2 s\upsilon \end{aligned} \tag{4.50}$$

Huynen's decomposition was greeted with comprehension, steadily but slowly so are his phenomenological argumentations. We note that its uniqueness is not guaranteed, because it is not basis-independent as was shown by Pottier [126].

4.7 Optimization of the Kennaugh matrix [K] for the Coherent Case Using Stokes' Vector Formulism

Using the Lagrange multiplier method applied to the received power matrices for the mono-static reciprocal case in terms of the Kennaugh (Stokes reflection) matrices $[K_c]$, $[K_x]$, and $[K_m]$ of (3.30) and (3.31), respectively, derived in [113], enables one to determine the characteristic polarization states similar to the generalized ρ-transformation methods.

For simplicity, the transmitted wave incident on the scatterer is assumed to be a completely polarized and normalized wave $\mathbf{q}'$ so that

$$q'_o = (q_1'^2 + q_2'^2 + q_3'^2)^{1/2} = 1 \tag{4.51}$$

and re-stating the received power expression defined in (3.26), (3.29), and (3.34) as functions of the optimal Stokes parameters, where

$$P_c = q'^T [K_c] q' = P_c(q_1', q_2', q_3'), \quad P_x = q'^T [K_x] q' = P_x(q_1', q_2', q_3'), \\ P_m = q'^T [K_m] q' = P_m(q_1', q_2', q_3') \tag{4.52}$$

are subject to the constraint of (4.51).This requirement dictates the use of the method of Lagrangian multipliers to find the extrema of the received powers P_c, P_x, and P_m. Reformulating
the equation of constraint to be given by

$$\phi(q_1', q_2', q_3') = (q_1'^2 + q_2'^2 + q_3'^2)^{1/2} - 1 = 0 \tag{4.53}$$

then the Lagrangian multipliers method for finding the extreme values of any of the three returned power expressions $P_l(q_1', q_2', q_3')$ results with $l = c, m, x$ in

$$\frac{\partial P_e}{\partial q_i'} - \mu \frac{\partial \phi}{\partial q_i'} = 0 \qquad i = 1, 2, 3 \tag{4.54}$$

For the corresponding '*degenerate deterministic*' (purely coherent) Kennaugh matrix $[K]$, insertion of the corresponding $P_l(q_1', q_2', q_3')$ into (4.54) results in a set of Galois equations yielding for:

(i) the extreme *'co-polar channel power* P_c*'* four solutions: *two maxima* $\rho_{cm1,2}$ which are orthogonal, and *two minima (nulls)* which may in an extreme pathological case be orthogonal (or identical but generally are not); and not one more or not one less of any of these extrema;

(ii) the returned *'cross-polar channel power* P_x*'* with six extreme solutions, being the three non-identical pairs of orthogonal polarization states: the *cross-polar maxima* $\rho_{xm1,2}$, the *cross-polar minima* $\rho_{xn1,2}$, and the *cross-polar saddle optima* $\rho_{xs1,2}$; and not one more or not one less of any of these extrema;

(iii) the returned power for the *'matched antenna case* P_m*'* yields only two solutions being identical to the *'co-polar maxima pair* $\rho_{cm1,2} = \rho_{xn1,2}$' and not one more or not one less of any of these extrema.

In summary, $Opt\{P_c(q_1', q_2', q_3')\}$ yields always exactly four solutions; $Opt\{P_x(q_1', q_2', q_3')\}$ yields always exactly six solutions (or equivalently three "bi-orthogonal" pairs of orthogonal polarization states); and $Opt\{P_m(q_1', q_2', q_3')\}$ yields always two solutions (or equivalently, exactly one pair of orthogonal polarization states). This represents an important result which also holds for the partially polarized case subject to incidence of a completely polarized wave.

A comparison of methods for a coherent scatterer of a specifically given Sinclair matrix $[S]$, with corresponding $[G]$ and $[K]$, plus $[K_c]$, $[K_x]$, and $[K_m]$, analyzed in [113], clearly demonstrates that all of the methods introduced are equivalent.

4.8 Determination of the Polarization Density Plots by van Zyl, and the Polarization Phase Correlation Plots by Agrawal

In radar meteorology, and especially in *'Polarimetric Doppler Radar Meteorology'* the Kennaugh target matrix characteristic operator concept was well received and further developed in the thesis of Agrawal [78]; and especially analyzed in depth by McCormick [127], and Antar [128] because various hydro-meteoric parameters can directly be associated with the Huynen or alternate McCormick parameters. In radar meteorology, the Poincaré sphere visualization of the characteristic polarization states has become commonplace; whereas in wide area SAR remote sensing the co/cross-polarization and Stokes parameter power density plots on the unwrapped planar transformation of the polarization sphere surface, such as introduced independently - *at the same time* - in the dissertations of van Zyl [79] and Agrawal [78], are preferred.

Because of the frequent use of the *'co/cross-polarization power density plots'*, $P_c(\rho), P_{c\perp}(\rho)$ and $P_x(\rho)$; and the equally important but hitherto rarely implemented *'co/cross-polarization phase correlation plots* $R_c(\rho), R_{c\perp}(\rho)$ *and* $R_x(\rho), R_{x\perp}(\rho)$ *'*, those are here introduced. Following Agrawal [78], who first established the relation between the *'Scattering Matrix Characteristic Operators of Kennaugh and Huynen'* with the *'polarimetric power-density/ phase-correlation plots'*, we obtain for the reciprocal case $S_{AB} = S_{BA}$

$$[C_{3L}] = \langle \mathbf{f}_{3L} \cdot \mathbf{f}_{3L}^{\dagger} \rangle = \begin{bmatrix} \langle |S_{AA}|^2 \rangle & \sqrt{2}\langle S_{AA}S_{AB}^* \rangle & \langle S_{AA}S_{BB}^* \rangle \\ \sqrt{2}\langle S_{AB}S_{AA}^* \rangle & 2\langle |S_{AB}|^2 \rangle & \sqrt{2}\langle S_{AB}S_{BB}^* \rangle \\ \langle S_{BB}S_{AA}^* \rangle & \sqrt{2}\langle S_{BB}S_{AB}^* \rangle & \langle |S_{BB}|^2 \rangle \end{bmatrix} \tag{4.55}$$

re-expressed in terms of the co/cross-polarimetric power density expressions:

$$P_c(\rho) = \langle |S_{AA}|^2 \rangle \qquad P_{c\perp}(\rho) = \langle |S_{BB}|^2 \rangle \qquad P_x(\rho) = \langle |S_{AB}|^2 \rangle = P_{x\perp}(\rho) \tag{4.56}$$

and the co/cross-polarization phase correlation expressions:

$$R_c(\rho) = \langle S_{AA}S_{BB}^* \rangle \qquad R_{c\perp}(\rho) = \langle S_{BB}S_{AA}^* \rangle \qquad R_x(\rho) = \langle S_{AA}S_{AB}^* \rangle \qquad R_{x\perp}(\rho) = \langle S_{BB}S_{AB}^* \rangle \tag{4.57}$$

so that $[C_{L3}(\rho)]$ may be rewritten according to (3.68) – (3.72) as

$$[C_{3L}(\rho)] = \begin{bmatrix} P_c(\rho) & \sqrt{2}\,R_x(\rho) & R_c(\rho) \\ \sqrt{2}\,R_x(\rho)^* & 2P_x(\rho) & \sqrt{2}\,R_x^{\perp}(\rho)^* \\ R_c(\rho)^* & \sqrt{2}\,R_x^{\perp}(\rho) & P_c^{\perp}(\rho) \end{bmatrix} \tag{4.58}$$

satisfying according to (3.70) and (3.71) the following inter-channel and symmetry relations

$$P_c\left(\rho_{\perp} = -\frac{1}{\rho^*}\right) = P_c^{\perp}(\rho) \qquad \left|R_x\left(\rho_{\perp} = -\frac{1}{\rho^*}\right)\right| = |R_x^{\perp}(\rho)|$$

$$P_x\left(\rho_{\perp} = -\frac{1}{\rho^*}\right) = P_x(\rho) \qquad \left|R_c\left(\rho_{\perp} = -\frac{1}{\rho^*}\right)\right| = |R_c(\rho)| \tag{4.59}$$

and frequently also the degree of polarization $D_p(\rho)$ and the degree of coherency $\mu(\rho)$ in terms of the directly measurable $P_c(\rho), P_{c\perp}(\rho)$ and $P_x(\rho)$; $R_c(\rho), R_{c\perp}(\rho)$ and $R_x(\rho), R_{x\perp}(\rho)$, provided a 'dual-orthogonal, dual-channel measurement system for coherent and partially coherent scattering ensembles is available requiring high-resolution, high channel isolation, high side-lobe reduction, and high sensitivity polarimetric amplitude and phase correlation, where

$$\mu(\rho)=\frac{|R_x(\rho)|}{\sqrt{P_c(\rho)P_x(\rho)}},\quad D_p(\rho)=\frac{\left\{[P_c(\rho)-P_x(\rho)]^2+4|R_x(\rho)|^2\right\}^{1/2}}{(P_c(\rho)+P_x(\rho))},\quad \textit{where}\ \ 0\le\mu(\rho)\le D_p(\rho)\le 1 \tag{4.60}$$

and for coherent (deterministic) scatterers $\mu = D_p = 1$, whereas for completely depolarized scatterers $\mu = D_p = 0$.

The respective power-density profiles and phase-correlation plots are then obtained from the normalized polarimetric radar brightness functions as functions of (ϕ,τ) with $\frac{-\pi}{2}\le\phi\le\frac{\pi}{2}$, $\frac{-\pi}{4}\le\tau\le\frac{\pi}{4}$ so that

$$\begin{aligned}
V_{AA}(\phi,\tau)&=\mathbf{p}^T(\phi,\tau)[S(HV)]\mathbf{p}(\phi,\tau)\\
V_{AB}(\phi,\tau)&=\mathbf{p}_\perp^\dagger(\phi,\tau)[S(HV)]\mathbf{p}(\phi,\tau)\\
V_{BA}(\phi,\tau)&=\mathbf{p}^T(\phi,\tau)[S(HV)]\mathbf{p}_\perp(\phi,\tau)\\
V_{BB}(\phi,\tau)&=\mathbf{p}_\perp^\dagger(\phi,\tau)[S(HV)]\mathbf{p}_\perp(\phi,\tau)
\end{aligned} \tag{4.61}$$

where

$$\begin{aligned}
P_c&=|V_{AA}|^2=|S_{AA}(\phi,\tau)|^2\\
P_x&=|V_{AB}|^2=|S_{AB}(\phi,\tau)|^2\\
R_c&=|\phi_{AA}-\phi_{BB}|=|\arg V_{AA}(\phi,\tau)-\arg V_{BB}(\phi,\tau)|\\
R_x(\phi,\tau)&=|\phi_{AA}-\phi_{AB}|=|\arg V_{AA}(\phi,\tau)-\arg V_{AB}(\phi,\tau)|\\
R_{x\perp}(\phi,\tau)&=|\phi_{BB}-\phi_{BA}|=|\arg V_{BB}(\phi,\tau)-\arg V_{BA}(\phi,\tau)|
\end{aligned} \tag{4.62}$$

In addition, the Maximum Stokes Vector $\mathbf{q}_{0MAX}$, and the maximum received power density P_m may be obtained from

$$P_m(\phi,\tau)=\mathbf{q}_0(\phi,\tau)=[K]\mathbf{q}(\phi,\tau) \tag{4.63}$$

where examples are provided in Figs. 4.5 and 4.6 for one specific matrix case [31, 82] given by

$$[S(HV)]=\begin{bmatrix}2j & 0.5\\ 0.5 & -j\end{bmatrix} \tag{4.64}$$

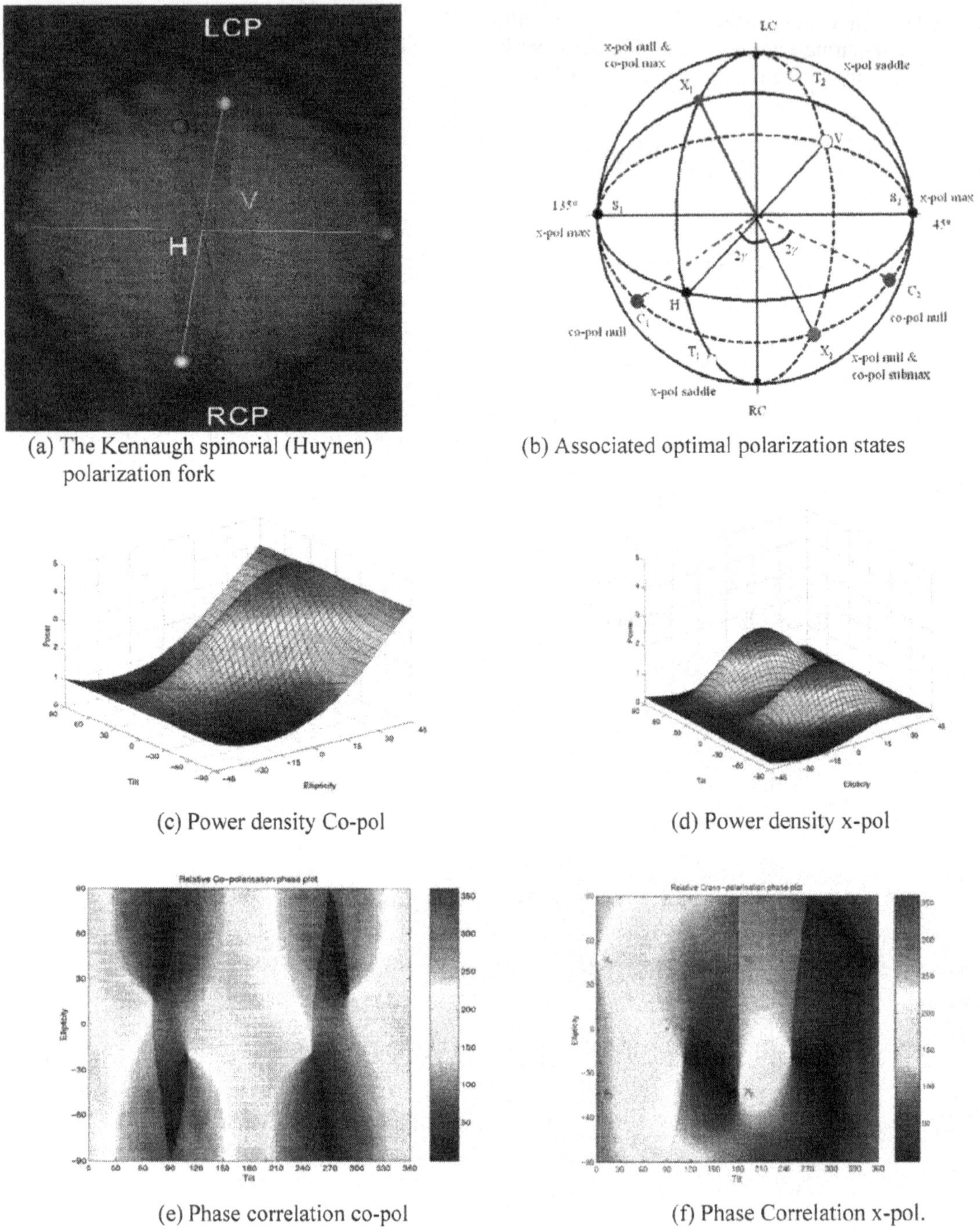

(a) The Kennaugh spinorial (Huynen) polarization fork

(b) Associated optimal polarization states

(c) Power density Co-pol

(d) Power density x-pol

(e) Phase correlation co-pol

(f) Phase Correlation x-pol.

Fig. 4.5 Power Density and Phase Correlation Plots for eq. 4.64 (by courtesy of James Morris)

4.9 Optimal Polarization States and its Correspondence to the Density Plots for the Partially Polarized Cases

According to the wave dichotomy portrayed for partially polarized waves, there exists one case for which the coherency matrix for the partially polarized case may be separated into one fully polarized and one completely depolarized component vector according to Chandrasekhar [34]. This principle will here be loosely applied to the case for which a completely polarized wave is incident on either a temporally incoherent (e.g., hydro-meteoric scatter) or spatially incoherent (e.g., rough surface viewed from different depression angles as in synthetic aperture radar imaging). This allows us to obtain a first order approximation for dealing with partially coherent and/or partially polarized waves when the polarimetric entropy is low; for which we then obtain the following optimization criteria:

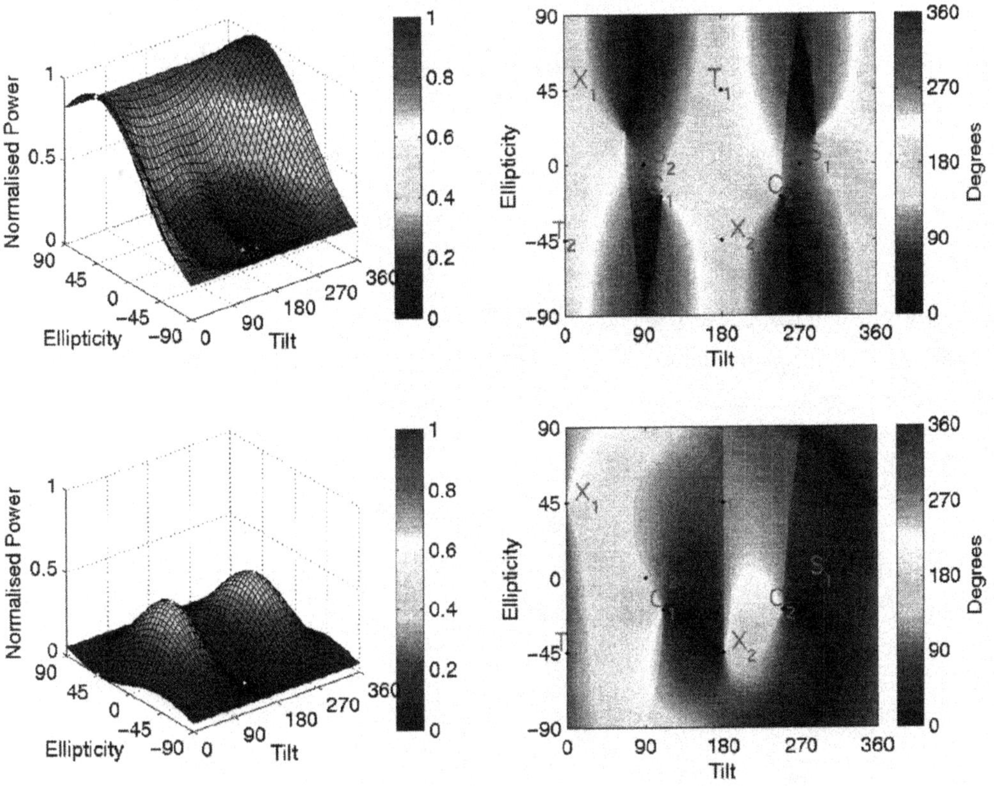

Fig. 4.6 Power Density and Phase Correlation Plots for eq. 4.64 (by courtesy of James Morris)

The energy density arriving at the receiver back-scattered from a distant scatterer ensemble subject to a completely polarized incident wave may be separated into four distinct categories where the Stokes vector is here redefined with $\mathbf{q}_p$ and $\mathbf{q}_u$ denoting the completely polarized and the unpolarized components, respectively

$$\mathbf{q} = \mathbf{q_p} + \mathbf{q_u} = \begin{bmatrix} q_0 \\ q_1 \\ q_2 \\ q_3 \end{bmatrix} = \begin{bmatrix} D_p q_0 \\ q_1 \\ q_2 \\ q_3 \end{bmatrix} + \begin{bmatrix} (1-D_p)q_0 \\ 0 \\ 0 \\ 0 \end{bmatrix} \tag{4.65}$$

and $\mathbf{q}$ as well as D_p were earlier defined, so that the following four categories for optimization of partially polarized waves can be defined as

q_0 total energy density in the scattered wave before it reaches the receiver

$q_0 D_p$ completely polarized part of the intensity

$q_0(1-D_p)$ noise of the unpolarized part

(4.66)

$\frac{1}{2}q_0(1+D_p)$ maximum of the total receptable intensity, the sum of the matched polarized part and one half of the unpolarized part: $\{D_p q_0\}+\left\{\frac{1}{2}(1-D_p)q_0\right\}=\left\{\frac{1}{2}(1+D_p)q_0\right\}$

Considering a time-dependent scatterer which is illuminated by a monochromatic (completely polarized wave) $\mathbf{E}^t$, for which the reflected wave $\mathbf{E}^s$ is, in general, non-monochromatic; and therefore, partially polarized. Consequently, the Stokes vector and Kennaugh matrix formulism will be applied to the four types of energy density terms defined above in (4.66).

4.10 Optimization of the Adjustable Intensity $D_p q_0$

The energy density $D_p q_0$, contained in the completely polarized part $\mathbf{q}_p$ of $\mathbf{q}$, is called the adjustable intensity because one may adjust the polarization state of the receiver to ensure the polarization match as shown previously for the coherent case. We may rewrite the scattering process in index notation as

$$q_i^s=\sum_{j=0}^{3}K_{ij}q_j^t \qquad where \quad j=0,1,2,3 \tag{4.67}$$

The adjustable intensity $D_p q_0$ can be re-expressed as

$$D_p q_0^s=\left(\sum_{i=0}^{3}q_i^{s2}\right)^{1/2}=\left[\sum_{i=1}^{3}\left(\sum_{j=0}^{3}K_{ij}q_j^t\right)^2\right]^{1/2} \tag{4.68}$$

where the q_i^t are the elements of the Stokes vector of the transmitted wave. The partial derivative of $(D_p q_0)^2$ with respect to q_k^t can be derived as

$$\frac{\partial(D_p q_0^s)^2}{\partial q_k^t}=\sum_{i=1}^{3}\frac{\partial q_i^{s2}}{\partial q_k^t}=2\sum_{i=1}^{3}q_i^s K_{ik}=2\sum_{i=1}^{3}\sum_{j=0}^{3}K_{ij}K_{ik}q_j^t \tag{4.69}$$

For optimizing the adjustable intensity, we apply the method of Lagrangian multipliers, which yields

$$\frac{\partial(D_p q_0^s)^2}{\partial q_k^t}-\mu\frac{\partial\phi}{\partial q_k^t}=2\sum_{i=1}^{3}\sum_{j=0}^{3}K_{ij}K_{ik}q_j^t-\mu q_k^t \tag{4.70}$$

where ϕ is the constraint equation

$$\phi(q_1^t,q_2^t,q_3^t)=(q_1^{t2},q_2^{t2},q_3^{t2})^{1/2}-1=0 \tag{4.71}$$

Equation (4.70) subject to (4.71) constitutes a set of inhomogeneous linear equations in $q_1^t(\mu)$, $q_2^t(\mu)$ *and* $q_3^t(\mu)$, with solutions as three functions of μ. Substituting $q_i^t(\mu; i=1,2,3)$ into the constrained condition (4.71) leads to a sixth-order polynomial Galois equation of μ. For each μ value,

q_1^t, q_2^t, q_3^t, *and* $D_p q_0^s$ are calculated according to (4.67) to (4.69). The largest (or smallest) intensity is the optimal intensity; the corresponding $\mathbf{q}^t$ is the optimal polarization state of the transmitted wave.

4.11 Minimizing the Noise-Like Energy Density Term: $q_0^s(1-D_p)$

An unpolarized wave can always be represented by an incoherent sum of any two orthogonal completely polarized waves of equal intensity [14, 15],which leads to 50% efficiency for the reception of the unpolarized wave. In order to receive as much 'polarized energy' as possible, the noise-like energy needs to be minimized. The total energy density of the unpolarized part of the scattered wave is given by:

$$\left(1-D_p\right)q_0^s = q_0^s - D_p q_0^s = \sum_{j=0}^{3} K_{0j} q_j^t - \sqrt{\sum_{i=1}^{3}\left(\sum_{j=0}^{3} K_{ij} q_j^t\right)^2} \tag{4.72}$$

Hitherto, no simple method was found for finding the analytic closed form solution for the minimum; instead, numerical solutions have been developed and are in use.

4.12 Maximizing the Receivable Intensity in the Scattered Wave: $\frac{1}{2}q_0(1+D_p)$

The total receivable energy density consists of two component parts: 100% reception efficiency for the completely polarized part of the scattered wave and 50% reception efficiency for the unpolarized part. The resulting expression for the total receivable intensity:

$$\frac{1}{2}\left(1+D_p\right)q_0^s = D_p q_0^s + \frac{1}{2}\left(1-D_p\right)q_0^s = \frac{1}{2}\sum_{j=0}^{3} K_{0j} q_j^t + \frac{1}{2}\sqrt{\left(\sum_{i=1}^{3}\sum_{j=0}^{3} K_{ij} q_j^t\right)^2} \tag{4.73}$$

can only be solved using numerical analysis and computation. The resulting maximally received Stokes vector is plotted in Fig. 4.7 (where q was replaced by p); and we observe that for the fully polarized case no '*depolarization pedestal*' exists. It appears as soon as $p<1$, and for $p=0$ it reaches its maximum of 0.5 for which the polarization diversity profile has deteriorated into the '*flat equal power density profile*', stating that the '*polarization diversity*' becomes meaningless.

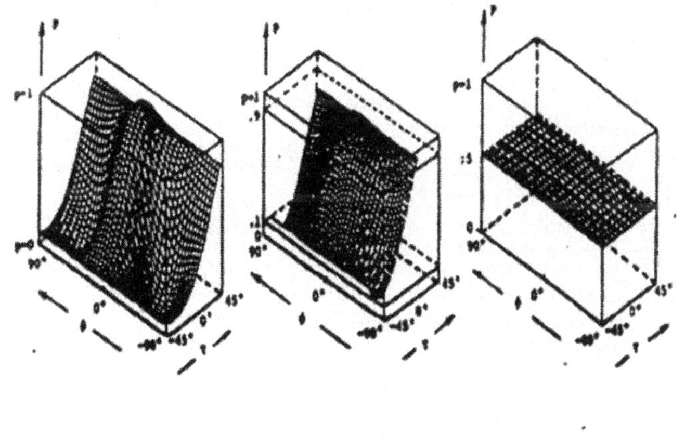

p=1
coherent
point scatterer

p = .8
distributed partially
coherent scatterer

p = 0
total polarization
noise

Fig. 4.7 Optimal Polarization States for the Partially Polarized Case

In conclusion, we refer to Boerner et al. [31, 82], where an optimization procedure for (4.65 – 4.68) in terms of $[K]$ for a completely polarized incident wave is presented together with numerical examples. It should be noticed here that Yang more recently provided another more elegant method in [119 - 121] for analyzing the statistical optimization procedure of the Kennaugh matrix.

5. References

1. Boerner, W-M., "Recent advances in extra-wide-band polarimetry, interferometry and polarimetric interferometry in synthetic aperture remote sensing, and its applications," *IEE Proc.-Radar Sonar Navigation, Special Issue of the EUSAR-02,* vol. 150, no. 3, June 2003, pp. 113-125
2. Sinclair, G., "Modification of the Radar Target Equation for Arbitrary Targets and Arbitrary Polarization", Report 302-19, Antenna Laboratory, The Ohio State University Research Foundation, 1948.
3. Sinclair, G., "The Transmission and Reception of Elliptically Polarized Waves", *Proceedings of the IRE*, vol. 38, no. 2, pp. 148-151, 1950.
4. Kennaugh, E. M., "Polarization Properties of Radar Reflections", Master's thesis, Ohio State University, Columbus, March 1952.
5. Kennaugh, E.M., "Effects of the Type of Polarization on Echo Characteristics", Reports 381-1 to 394-24, Antenna Laboratory, The Ohio State University Research Foundation, 1949-1954.
6. Deschamps, G.A., "Geometrical Representation of the Polarization of a Plane Electromagnetic Wave", *Proceedings of the IRE*, vol.39, no.5, pp. 540-544, 1951.
7. Graves, C.D., "Radar Polarization Power Scattering Matrix", *Proceedings of the IRE*, vol. 44, no. 5, pp. 248-252, 1956.
8. Copeland, J.R., "Radar Target Classification by Polarization Properties", *Proceedings of the IRE*, vol. 48, no. 7, pp. 1290-1296, 1960.
9. Huynen, J.R., "Phenomenological Theory of Radar Targets", Ph.D. thesis, University of Technology, Delft, The Netherlands, December 1970.
10. Poelman, A.J. and J.R.F. Guy, 1985, "Polarization information utilization in primary radar"; in Boerner, W-M. et al. (eds), 1985, *"Inverse Methods in Electromagnetic Imaging"*, Proceedings of the NATO-Advanced Research Workshop, (18-24 Sept. 1983, Bad Windsheim, FR Germany), Parts 1&2, NATO-ASI C-143, (1,500 pages), D. Reidel Publ. Co., Part 1, pp. 521-572.
11. Poelman, A.J. and K.J. Hilgers, 1988, "Effectiveness of multi-notch logic-product polarization filters for countering rain clutter", in *Proceedings of the NATO Advanced Research Workshop on Direct and Inverse Methods in Radar Polarimetry*, W.-M. Boerner et al (eds), Bad Windsheim, Germany, September 18-24, 1988; Kluwer Academic Publishers, Dordrecht 1992; NATO ASI Series C-350, Part 2, pp. 1309-1334.
12. Root, L.W., ed., 1982, *Proceedings of the First Workshop on Polarimetric Radar Technology*, GACIAC, Chicago, IL: 295 p. See also Proceedings of the second and third workshops, 1983 and 1989 respectively, with the same publisher.
13. Boerner, W-M, and M.B. El-Arini, "Polarization Dependence in Electromagnetic Inverse Problem", *IEEE Transactions on Antennas and Propagation*, vol.29, no. 2, pp. 262-271, 1981.
14. Boerner, W-M, (ed.), *Direct and Inverse Methods in Radar Polarimetry*, NATO ASI Series C, Math. and Phys. Science, Kluwer Academic Publishers, Netherlands, 1985.
15. Boerner, W-M, (ed.), *Inverse Methods in Electromagnetic Imaging*, NATO ASI Series C, Math. and Phys. Science, Kluwer Academic Publishers, Netherlands, 1992.
16. Boerner, W-M., "Use of Polarization in Electromagnetic Inverse Scattering", Radio Science, Vol. 16(6) (Special Issue: 1980 Munich Symposium on EM Waves), pp. 1037-1045, Nov./Dec. (1981b).
17. Boerner, W-M, "Polarimetry in Remote Sensing and Imaging of Terrestrial and Planetary Environments", Proceedings of Third International Workshop on Radar Polarimetry (JIPR-3, 95), IRESTE, Univ.-Nantes, France, pp. 1-38, 1995b.
18. Boerner, W-M., 'Recent Advances in Polarimetric Interferometric SAR Theory & Technology and its Application' ESA-CEOS-MRS'99, SAR Cal-Val-Workshop, CNES, Toulouse, FR, 1999 Oct. 25-29

19. Boerner, W-M, H. Mott, E. Lüneburg, C. Livingston, B. Brisco, R. J. Brown and J. S. Paterson with contributions by S.R. Cloude, E. Krogager, J. S. Lee, D. L. Schuler, J. J. van Zyl, D. Randall P. Budkewitsch and E. Pottier, "Polarimetry in Radar Remote Sensing: Basic and Applied Concepts", Chapter 5 in F.M. Henderson, and A.J. Lewis, (eds.), Principles and Applications of Imaging Radar, Vol. 2 of Manual of Remote Sesning, (ed. R.A. Reyerson), 3rd Ed., John Willey & Sons, New York, 1998.
20. Boerner, W-M, S.R. Cloude, and A. Moreira, 2002, "User collision in sharing of electromagnetic spectrum: Frequency allocation, RF interference reduction and RF security threat mitigation in radio propagation and passive and active remote sensing", *URSI-F Open Symp.*, Sess. 3AP-1, 2002 Feb. 14, Garmisch-Partenkirchen, Germany.
21. Azzam, R. M. A. and N. M. Bashara, 1977, *Ellipsometry and Polarized Light*, North Holland, Amsterdam: 539 p.
22. Beckmann, P., 1968. *The Depolarization of Electromagnetic Waves*, The Golem Press, Boulder, CO: 214 p.
23. Chipman, R. A., and J. W. Morris, eds., "Polarimetry: Radar, Infrared, Visible, Ultraviolet, X-Ray", Proc. SPIE-1317, 1990 (also see SPIE Proc. 891, 1166, 1746, 1988, 1989, and 3121).
24. Jones, R., 1941, "A new calculus for the treatment of optical systems, I. Description and discussion", J. Opt. Soc. Am., 31 (July 1941), pp. 488-493; "II. Proof of the three general equivalence theorems, ibid. pp. 493-499; III. The Stokes theory of optical activity", ibid. pp. 500-503; ibid. 32 (1941), pp. 486-493 , ibid. 37 (1947), pp. 107-110 (See also Swindell, W., 1975, *Polarized Light*, Halsted Press/John Wiley & Sons, Stroudsburg, PA: pp. 186-240).
25. Boerner, W-M, C. L. Liu, and Zhang, "Comparison of Optimization Processing for 2x2 Sinclair, 2x2 Graves, 3x3 Covariance, and 4x4 Mueller (Symmetric) Matrices in Coherent Radar Polarimetry and its Application to Target Versus Background Discrimination in Microwave Remote Sensing", EARSeL Advances in Remote Sensing, Vol. 2(1), pp. 55-82, 1993.
26. Cloude S.R., "Polarimetry in Wave Scattering Applications", Chapter 1.6.2 in SCATTERING, Eds R Pike, P Sabatier, Academic Press, to be published December 1999
27. Cloude, S.R. and E. Pottier, "A review of target decomposition theorems in radar polarimetry", IEEE Trans. GRS, vol. 34(2), pp. 498-518, Mar. 1996.
28. Cloude, S. R., 1992, Uniqueness of Target Decomposition Theorems in Radar Polarimetry. *Direct and Inverse Methods in Radar Polarimetry*, Part 1, Boerner, W-M, ed. Kluwer Academic Publishers, Dordrecht, The Netherlands: 267-296.
29. Cloude, S.R., Polarimetry: The Characterization of Polarimetric Effects in EM Scattering, Ph.D. thesis, University of Birmingham, Faculty of Engineering, Birmingham, England/UK, Oct. 1986.
30. Krogager, E., Decomposition of the Sinclair Matrix into Fundamental Components with Application to High Resolution Radar Target Imaging, *Direct and Inverse Methods in Radar Polarimetry*, Part 2, Boerner, W-M., ed. Kluwer Academic Publishers, Dordrecht, The Netherlands: 1459-1478, 1992.
31. Boerner, W-M. and A-Q. Xi, 1990, The Characteristic Radar Target Polarization State Theory for the Coherent Monostatic and Reciprocal Case Using the Generalized Polarization Transformation Ratio Formulation, *AEU*, 44 (6): X1-X8.
32. Bebbington, D.H.O, "The Expression of Reciprocity in Polarimetric Algebras", Progress in Electromagnetics Research Symposium (PIERS'98), *Procs. Fourth Int'l. Workshop on Radar Polarimetry (JIPR-98)*, pp. 9-18, IRESTE, Nantes, July 1998.
33. Czyz, Z.H., "Basic Theory of Radar Polarimetry - An Engineering Approach", Prace PIT, No.119, 1997, Warsaw, Poland, pp.15-24.
34. Born, M. and E. Wolf, *Principles of Optics*, 3rd ed. Pergamon Press, New York: 808 p., 1965.
35. Goldstein, D. H. and R. A. Chipman, "Optical Polarization: Measurement, Analysis, and Remote Sensing", Proc. SPIE-3121, 1997 (see Proc. SPIE 891, 1166, 1317, 1746, 1988, 1989: OPT-POL).
36. Krogager, E. and W-M. Boerner, On the importance of utilizing complete polarimetric information in radar imaging and classification, AGARD Symposium: Remote Sensing - A Valuable Source of Information, Toulouse, France, 1996 April 22-25, *AGARD Proc.*, (528 pp.), pp. 17.1 - 17.12.
37. Krogager, E., Aspects of Polarimetric Radar Imaging, Ph.D. thesis, Technical University of Denmark (TUD), Electromagnetics Institute, Lyngby, DK, March 1993.

38. Krogager, E., Comparison of Various POL-RAD and POL-SAR Image Feature Sorting and Classification Algorithms, Journées Internationales de la Polarimétrie Radar, Proc. JIPR'98, pp. 77-86, Nantes, France, 13-17 July, 1998.
39. Krogager, E., and Z.H. Czyz, "Properties of the Sphere, Di-plane and Helix Decomposition" Proc. of 3rd International Workshop on Radar Polarimetry, IRESTE, University of Nantes, France, pp. 106-114, April 1995.
40. Krogager, E., W.-M. Boerner, S.N. Madsen; "Feature-Motivated Sinclair Matrix (sphere/di-plane/helix) Decomposition and Its Application to Target Sorting For Land Feature Classification," SPIE-3120, 144-154, 1997..
41. Lee, J. S, M R Grunes, T L Ainsworth, L J Du, D L Schuler, S R Cloude, "Unsupervised Classification using Polarimetric Decomposition and the Complex Wishart Distribution", IEEE Transactions Geoscience and Remote Sensing, Vol 37/1, No. 5, p 2249-2259, September 1999
42. Lee, J.S. and M.R. Grunes, Polarimetric SAR Speckle Filtering and Terrain Classification-An Overview 26p., Book Ch. in XX, World Sci. Publ., Singapore, 1999
43. Lee, J.S., "Speckle suppression and analysis for synthetic aperture radar images", *SPIE Optical Engineering*, Vol. 25 No. 5, pp. 636-643, May 1986.
44. Lee, J-S., M.R. Grunes and R. Kwok, Classification of multi-look polarimetric SAR imaging based on complex Wishart distributions, *Int'l Journal of Remote Sensing*, Vol. 15(11), pp. 2299-2311, 1994.
45. Boerner W-M., *'Report on POL & POL-IN Session'*, ESA-CEOS-MRS'99, SAR Cal-Val-Workshop., CNES Toulouse, FR, 1999 Oct. 29.
46. Tragl, K. "Polarimetric Radar Back-scattering from Reciprocal Random Targets", IEEE Trans. GRS-28(5), pp. 856 - 864, Sept. 1990 (see Dr. -Ing. Thesis, 1989).
47. Novak L.M., S. D. Halversen, G. J. Owirka, M. Hiett, "Effects of Polarization and Resolution on SAR ATR," IEEE Trans. AES, Vol. 33(1), pp. 102-116, Jan. 1997.
48. Novak, L.M., and C.M. Netishen, "Polarimetric Synthetic Aperture Radar Imaging", *Int'l Journal of Imaging Systems and Technology*, John Wiley & Sons, New York, NY, vol. 4, pp. 306-318, 1992.
49. Novak, L.M., and M.C. Burl, "Optimal Speckle Reduction in Polarimetric SAR Imagery" *IEEE Trans. AES*, vol. 26, no.2, pp. 293-305, 1990.
50. Novak, L.M., and M.C. Burl, and W.W. Irving, "Optimal Polarimetric Processing for Enhanced Target Detection", *IEEE Trans. AES*, vol. 29, no.1, pp. 234-244, 1993.
51. Novak, L.M., and S.R. Hesse, "Optimal Polarizations for Radar Detection and Recognition of Targets in Clutter", *Proceedings, 1993 IEEE National Radar Conference*, Lynnfield, MA, April 20-22, 1993, pp. 79-83.
52. Lüneburg, E., Radar polarimetry: A revision of basic concepts, in "Direct and Inverse Electromagnetic Scattering", H. Serbest and S. Cloude, eds., Pittman Research Notes in Mathematics Series 361, Addison Wesley Longman, Harlow, U.K., 1996, pp. 257 – 275.
53. Lüneburg, E., "Principles of Radar Polarimetry", Proceedings of the IEICE Trans. on the Electronic Theory, Vol. E78-C, no. 10, pp. 1339-1345, 1995 (see also: Lüneburg, E., Polarimetric target matrix decompositions and the 'Karhunen-Loeve expansion', *IGARSS'99, Hamburg, Germany*, June 28-July 2, 1999).
54. Lüneburg, E., Comments on "The Specular Null Polarization Theory" IEEE Trans. Geoscience and Remote Sensing, Vol. 35, 1997, pp. 1070 – 1071.
55. Lüneburg, E., V. Ziegler, A. Schroth, and K. Tragl, "Polarimetric Covariance Matrix Analysis of Random Radar Targets", pp.27.1 - 27.12, in Proc. NATO-AGARD-EPP Symposium on Target and Clutter Scattering and Their Effects on Military Radar Performance, Ottawa.Canada,1991 May 6 – 10 (also see: Lüneburg, E., M. Chandra, and W.-M. Boerner, Random target approximations, *Proc. PIERS Progress in Electromagnetics Research Symposium, Noordwijk, The Netherlands, July 11-15, 1994, CD Kluwer Publishers*, 1366-1369).
56. Cloude, S.R., Potential New Applications of Polarimetric Radar / SAR Interferometry, Proc. On Advances in Radar methods - with a view towards the Twenty First Century, EC-JRC/SAI (ISPRA) Hotel Dino, Baveno, Italy, 1998 July 20-22, (Proc. to be published in Fall 98)
57. Cloude S.R. and E. Pottier, "An Entropy-Based Classification Scheme for Land Applications of Polarimetric SAR", IEEE Trans GRS-35(1), 68-78, 1997

58. Kostinski, A.B., B.D. James, and W-M. Boerner, 1988, Polarimetric Matched Filter for Coherent Imaging, *Canadian Journal of Physics*, 66: 871-877.
59. Krogager, E., S.R. Cloude, J.-S. Lee, T.L. Ainsworth, and W.-M. Boerner, Interpretation of high resolution polarimetric SAR data, Journées Internationales de la Polarimétrie Radar, Proc. JIPR'98, pp. 165-170, Nantes, France, 13-17 July, 1998.
60. Kozlov, A. I., "Radar Contrast between Two Objects", *Radioelektronika*, 22 (7): 63-67, 1979.
61. Mott, H. and W-M. Boerner, editors, "Radar Polarimetry, SPIE's Annual Mtg., Polarimetry Conference Series", 1992 July 23 - 24, San Diego Convention Center, SPIE Conf. Proc. Vol. 1748, 1992
62. Boerner, W-M., B. Y. Foo, and H. J. Eom, 1987, Interpretation of the Polarimetric Co-Polarization Phase Term (φ_{HH} - φ_{VV}) in High Resolution SAR Imaging Using the JPL CV-990 Polarimetric L-Band SAR Data, Special IGARSS Issue of *IEEE Transactions on Geoscience and Remote Sensing*, 25 (1): 77-82.
63. Freeman, A. and S.T. Durden, "A Three-Component Scattering Model for Polarimetric SAR Data", IEEE Trans. GRS, Vol. 36(3), pp. 963-973, 1998.
64. Lee, J.S., M.R. Grunes and W.M. Boerner, "Polarimetric Property Preserving in SAR Speckle Filtering," Proceedings of SPIE, Vol. 3120, 236-242, San Diego, 1997.
65. van Zyl, J. J. "Application of Cloude's Target Decomposition Theorem to Polarimetric Imaging Radar Data", SPIE Proceedings (H. Mott, W-M Boerner eds.), Vol. 1748, San Diego, CA, 23-24 July 1992.
66. van Zyl, J. J., "An Overview of the Analysis of Multi-frequency Polarimetric SAR Data", Proceedings of the US-AU PACRIM Significant Results Workshop, MHPCC, Kihei, Maui, HI, 1999 August 24 - 26 (10 pages).
67. Schuler, D.L., J.S. Lee, G. De Grandi, "Measurement of Topography using Polarimetric SAR Images," *IEEE Trans. on Geoscience and Remote Sensing*, vol. 34, no.5, pp. 1266-1277, 1996.
68. Schuler, D.L, J.S. Lee, and T.L. Ainsworth, "Topographic mapping using polarimetric SAR data", *International Journal of Remote Sensing*, vol. 19(1), pp 141-160, 1988.
69. Schuler, D.L., J.S. Lee, T.L. Ainsworth, and M.R. Grunes, "Terrain topography measurement using multi-pass polarimetric synthetic aperture radar data," *Radio Science*, vol. 35, no.3, pp. 813-832, May-June 2000.
70. Bamler, R. and P. Hartl, "Synthetic Aperture Radar Interferometry", State of the Art Review, Inverse Problems, Vol. 14, pp. R1-R54, IOC Publications, Bristol, UK, 1998.
71. Zebker, H.A. and J.J. van Zyl, Imaging Radar Polarimetry: A Review, *Proceedings of the IEEE*, Vol. 79, pp. 1,583-1,606, 1991.
72. Madsen, S.N and H.A. Zebker, "Imaging Radar Interferometry," Chapter 6 (pp. 359-380) in *Manual of Remote Sensing*, Vol. 2, Principles and Applications of Imaging Radar, F. M. Henderson and A. J. Lewis, Eds., American Society for Photogrammetry and Remote Sensing, Wiley, New York, 940 p, 1998
73. Boerner, W-M., et al., (Guest Eds.), IEEE Transactions on the Antennas & Propagation Society, Vol. 29(2), Special Issue, Inverse Methods in Electromagnetics, (417 pages) 1980-81(1981a).
74. Dubois, P.C. and L. Norikane, 1987, "Data Volume Reduction for Imaging Radar Polarimetry", *Proceedings of IGARSS'87*, pp. 691-696, 1987.
75. Lee, J.S., K.P. Papathanassiou, T. L. Ainsworth, M.R. Grunes and A. Reigber, "A New Technique for Noise Filtering of SAR Interferometric Phase Images", IEEE Trans. GRS Vol. 36 , No.5 pp. 1456-1465.
76. Mott, H., Antennas *for Radar and Communications, A Polarimetric Approach*, John Wiley & Sons, New York, 1992, 521 p.
77. Chan, C-Y., *Studies on the Power Scattering Matrix of Radar Targets*, M.S. thesis, University of Illinois, Chicago, Illinois, 1981.
78. Agrawal, A. P., "A Polarimetric Rain Back-scattering Model Developed for Coherent Polarization Diversity Radar Applications", Ph.D. thesis, University of Illinois, Chicago, IL, December 1986.
79. van Zyl, J. J. "On the Importance of Polarization in Radar Scattering Problems", Ph.D. thesis, California Institute of Technology, Pasadena, CA, December 1985.
80. Papathanassiou, K. P., "Polarimetric SAR Interferometry", Ph.D. thesis, Tech. Univ. Graz, 1999.

81. Boerner, W-M., W-L. Yan, A-Q. Xi and Y. Yamaguchi, "On the Principles of Radar Polarimetry (Invited Review): The Target Characteristic Polarization State theory of Kennaugh, Huynen's Polarization Fork Concept, and Its Extension to the Partially Polarized Case", IEEE Proc., Special Issue on Electromagnetic Theory, Vol. 79(10), pp. 1538-1550, Oct. 1991.
82. Xi, A-Q and W-M. Boerner, "Determination of the Characteristic Polarization States of the target scattering matrix [S(AB)] for the coherent Monostatic and Reciprocal Propagation Space Using the Polarization Transformation Ratio Formulation", JOSA-A/2, 9(3), pp. 437-455, 1992.
83. van Zyl, J.J. and H. A. Zebker, 1990, Imaging Radar Polarimetry, *Polarimetric Remote Sensing, PIER 3*, Kong, J. A., ed. Elsevier, New York: 277-326.
84. van Zyl, J. J., H. Zebker, and C. Elachi, "Imaging Radar Polarization Signatures: Theory and Application", Radio Science, vol. 22, no. 4, pp. 529-543, 1987.
85. Pellat-Finet, P., "An introduction to a vectorial calculus for polarization optics", *Optik*, vol. 84 (5), pp. 169 – 175, 1990: Pellat-Finet, P., 1991, "Geometrical approach to polarization optics I: geometrical structure of polarized light", *Optik*, vol. 87, pp. 68-77.
86. Stratton, J.A., *Electromagnetic Theory*, McGraw-Hill, New York, 1941
87. Ulaby, F. T. and C. Elachi, Editors, Radar Polarimetry for Geo-science Applications, Artech House, Inc., Norwood, MA, 1990, 364p.
88. Lee, J.S., W-M Boerner, T. Ainsworth, D. Schuler, I. Hajnsek, and E. Lüneburg, "A Review of Polarimetric SAR Algorithms and its Applications", *Chinese Journal of Remote Sensing*, Chong-Li, Taiwan, in print, 2003.
89. Kostinski, A.B. and W.-M. Boerner, "On the polarimetric contrast optimization", *IEEE Trans. Antennas Propagt.*, vol. AP-35(8), pp. 988-991, 1987.
90. Agrawal, A. B. and W.-M. Boerner, "Redevelopment of Kennaugh's target characteristic polarization state theory using the polarization transformation ratio formalism for the coherent case", *IEEE Trans. Geosci. Remote Sensing*, AP-27(1), pp. 2-14, 1989.
91. Cameron, W.L., "Simulated polarimetric signatures of primitive geometrical shapes", *IEEE Trans. Geoscience Rem. Sens.*, 34, 3 793-803, 1996.
92. Pottier, E., *Contribution à la Polarimétrie Radar: de l'approche fondamentale aux applications*, Habilitation a Diriger des Recherches; Dr. Ing. habil., Ecole Doctorale Sciences pour l'Ingenieur de Nantes, Université de Nantes, l'IRESTE, La Chantrerie, Rue Pauc, BP 60601, F-44306 NANTES CED-3, 98-11-12, 1998 ; ibid., *"Radar Polarimetry : Towards a future Standardization"*, Annales des Télécommunications, vol 54 (1-2), pp 137-141, Janvier 1999.
93. Ferro-Famil, L.E., E. Pottier and J.S. Lee, "Unsupervised Classification of Multifrequency and Fully Polarimetric SAR Images Based on H/A/Alpha-Wishart Classifier," *IEEE Trans. GRS*, vol. 39(11), pp. 2332-2342, 2001.
94. Raney, R. K., "Processing Synthetic Aperture Radar Data", *Int'l JRS*, Vol. 3(3), pp. 243-257, 1982.
95. Cloude, S.R. and K. P. Papathanassiou, "Polarimetric optimization in radar interferometry", *Electronic Letters*, vol. 33(13), pp. 1176-1178, 1997.
96. Cloude, S.R. and K. P. Papathanassiou, Coherence optimization in polarimetric SAR interferometry, IGARSS'97Proc., Singapore, 1997 Aug. 03-09,Vol. IV, pp. 1932-1934, 1997.
97. Papathanassiou, K.P. and S.R. Cloude, "Single-baseline polarimetric SAR interferometry", *IEEE Trans. GRS*. 39 (6), pp. 2352-2363, 2001.
98. Papathanassiou, K.P. and J.R. Moreira, "Interferometric analysis of multi-frequency and multi-polarization SAR data", *Proc. IGARSS '96*, Lincoln, NE, Vol. III, pp. 1227-1229, 1996.
99. Reigber, A., *"Polarimetric SAR Tomography"*, Dissertation, University of Stuttgart, 2001 October 15 (ISSN 1434-8454, ISRN DLR-FB-2002-02), 2001.
100. Reigber A., A. Moreira, "First Demonstration of SAR Tomography using Polarimetric Airborne SAR Data". *IEEE Trans. GRS*, vol. 38 (5-1), pp 2142 -2152, 2000.
101. Reigber, A., A. Moreira and K.P. Papathanassiou, "First demonstration of airborne SAR tomography using multi-baseline data". *Proc. IGARSS-99* (06-28_07-02), A03_10:50, Hamburg, Germany, 1999.
102. IEEE 1983 *Standard Test Procedures for Antennas*, ANSI/IEEE-Std. 149-1979, IEEE-Publishing, ISBN 0-471-08032-2 (also see: Number 145-1983: Definitions of Terms for Antennas, *IEEE Transactions on Antennas and Propagation*, AP-31(6), November 1983, pp. II – 26).

103. Deschamps, G. A. and P. E. Mast, 1973, Poincaré Sphere Representation of Partially Polarized Fields, *IEEE Transactions on Antennas and Propagation*, 21 (4): 474-478.
104. Eaves, J. L. and E. K. Reedy, eds., 1987, *Principles of Modern Radar*, Van Nostrand Reinhold Company, New York: 712 p.
105. Cloude, S.R., E. Pottier, W-M Boerner, "Unsupervised Image Classification using the Entropy/Alpha/Anisotropy Method in Radar Polarimetry", *NASA-JPL, AIRSAR-02 Workshop*, Double Tree Hotel, Pasadena, CA, 2002 March 04-06.
106. Pottier, E., *"La Polarimétrie Radar Appliquée à la Télédétection"* Ecole Supérieure des Télécommunications de Tunis, Tunis, Tunisie, 1999 December 17.
107. Stokes, G.G., "On the composition and resolution of streams of polarized light from different sources", *Trans. Cambridge Philos. Soc.*, vol. 9, pp. 399-416, 1852 (also see: Stokes, G. G., 1901, Stokes's Mathematical and Physical Papers, Univ. Press, Cambridge).
108. Poincaré, H., *Théorie Mathématique de la Lumière*, Georges Carre, Paris, 310 p, 1892.
109. Zhivotovsky, L. A., Optimum Polarization of Radar Signals, *Radio Engineering and Electronic Physics*: 630-632, 1973, ibid. "The Polarization Sphere Modification to Four Dimensions for the Representation of Partially Polarized Electromagnetic Waves", Radiotechnica i Electronica, vol. 30, no. 8, pp. 1497-1504, 1985.
110. Czyz, Z.H., Czyz, Z.H., "Advances in the Theory of Radar Polarimetry," Prace PIT, No.117, Vol. XLVI, 1996, Warsaw, Poland, pp.21-28.
111. Misner, C.W., K.S. Thorne and A Wheeler, 1997, *Gravitation*, W.H. Freeman & Co., New York (twentieth printing: 1997)
112. Graham, A., *Kronecker Products and Matrix Calculus: with Applications*, New York: Ellis Horwood Ltd (John Wiley & Sons), 130 p, 1981.
113. Yan, W-L., et al, and W-M. Boerner, "Optimal polarization state determination of the Stokes reflection matrices [M] for the coherent case, and of the Mueller matrix [M] for the partially polarized case", *JEWA*, vol. 5(10), pp. 1123-1150, 1991.
114. Ziegler, V., E. Lüneburg, and A. Schroth, Mean Backscattering Properties of Random Radar Targets: A Polarimetric Covariance Matrix Concept, *Proceedings of IGARSS'92*, May 26-29, Houston Texas, pp. 266-268, 1992.
115. Priestley, M. B., *Spectral Analysis and Time Series, Volumes 1 & 2, Univariate and Multivariate Series, Prediction and Control*, Academic Press, New York, 1981.
116. Takagi, T., On an algebraic problem related to an analytical theorem of Caratheodory and Fejer and on an allied theorem of Landau, Japanese J. Math., 1, pp. 83-93, 1927
117. Horn, R. A. and Ch. R. Johnson, Topics in Matrix Analysis, Cambridge University Press, New York, 1991; ibid: Matrix Analysis, Cambridge University Press, New York, 1985
118. Kennaugh, E.M., "Polarization dependence of radar cross sections - A geometrical interpretation", *IEEE Trans. Antennas Propag.*, (Special Issue on Inverse Methods in Electromagnetic Scattering, W.-M. Boerner, A; K. Jordan, I. W. Kay, Gust Editors), vol. AP-29(3), pp. 412-414, 1981.
119. Yang, J., Y. Yamaguchi, H. Yamada, M. Sengoku, S. M. Lin, "Stable Decomposition of Mueller Matrix", IEICE Trans. Comm., vol. E81-B(6), pp. 1261-1268, June 1998
120. Yang, J., Yoshio Yamaguchi, Hiroyoshi Yamada, "Co-null of targets and co-null Abelian group," Electronics Letters, vol.35, no.12, pp.1017-1019, June 1999.
121. Yang, J., Yoshio Yamaguchi, Hiroyoshi Yamada, Masakazu Sengoku, Shi -Ming Lin, Optimal problem for contrast enhancement in polarimetric radar remote sensing, J-IEICE Trans. Commun., vol.E82-B, no.1, pp.174-183, Jan. 1999
122. Bogorodsky, V.V., D.B. Kanareykin, and A.I. Kozlov, "Polarization of the Scattered Radio Radiation of the Earth's Covers", Leningrad: Gidrometeorizdat, 1981 (in Russian).
123. Kanareykin, D.B., N.F. Pavlov, U.A. Potekhin, "The Polarization of Radar Signals", Moscow: Sovyetskoye Radio, Chap. 1-10 (in Russian), 1986, (English translation of Chapters 10-12: "Radar Polarization Effects", CCM Inf. Corp., G. Collier and McMillan, 900 Third Ave., New York, NY 10023).
124. Cameron, W.L. and L.K. Leung, "Feature-motivated scattering matrix decomposition", *Proc. IEEE Radar Conf.*, Arlington, VA, May 7-10, 1990, pp. 549-557, 1990.

125. Barnes, R.M., "Roll Invariant Decompositions for the Polarization Covariance Matrix", Internal Report, Lincoln Laboratory, MIT, Lex., MA 02173-0073.
126. Pottier, E., Contribution de la Polarimétrie dans la Discrimination de Cibles Radar, Application à l'Imagérie Electromagnétique haute Resolution, Ph.D. thesis, IRESTE, Nantes, France, December 1990.
127. McCormick, G. C., "The Theory of Polarization Diversity Systems (in Radar Meteorology): The partially polarized case", *IEEE Trans. Ant. & Prop.*, vol. AP-44(4), pp. 425-433, 1996. (McCormick, G. C. and A. Hendry, 1985, "Optimum polarizations for partially polarized backscatter", *IEEE Trans. Antennas Propagation*, AP-33(1), pp. 33-39).
128. Antar, Y.M.M., "Polarimetric Radar Applications to Meteorology", in W-M Boerner et al. (eds.) *Direct and Inverse Methods in Radar Polarimetry*, Part 2, pp. 1683-1695, Kluwer Academic Publishers, Dordrecht, NL, 1992.
129. Kostinski, A.B, and W-M. Boerner, "On foundations of radar polarimetry", *IEEE Trans. Antennas Propagation*, vol. AP-34, pp. 1395-1404, 1986; H. Mieras, "Comments on 'Foundations of radar polarimetry'", ibid, pp. 1470-1471; "Authors's reply to 'Comments' by H. Mieras", ibid. pp. 1471-1473.
130. Yamaguchi, Y., Y. Takayanagi, W-M. Boerner, H. J. Eom and M. Sengoku, "Polarimetric Enhancement in Radar Channel Imagery", IEICE Trans. Communications, vol. E78-B, no. 1, pp. 45-51, Jan 1996
131. Holm, W.A., R.M.. Barnes, "On radar polarization mixed target state decomposition techniques", *Proceedings of the 1988 National Radar Conference*, pp. 248-254, April 1988.
132. Ostrobityanov, R.V., and F.A. Basalov, "Statistical Theory of Distributed Targets", Dedham, MA: Artech House, 364 pages, 1985.
133. Potekhin, V.A. and V.N. Tatarinov, "The Theory of Coherence of Electromagnetic Fields", Moscow, M. Sov. Radio, 1978.
134. Touzi R., A. Lopes, J. Bruniquel , and P. Vachon, "Unbiased estimation of the coherence for SAR Imagery", IEEE Trans. Geoscience Remote Sensing, Vol. 37, No. 1, Jan. 1999
135. Rignot, E., R. Chellapa, and P. Dubois, Unsupervised segmentation of polarimetric SAR data using the covariance matrix, IEEE Trans. Geoscience and Remote Sensing, 30(4) 697-705, 1992.
136. Von Neumann, J., "Distribution of the ratio of the mean-square successive difference to the variance". *Ann. Math. Statist.*, 12, 367-395.
137. Goodman, N. R., Statistical analysis based on a certain multi-variate complex Gaussian distribution (an introduction), *Annals of Mathematics and Statistics*, 34 (1963) 152-177

6. Appendices

A. The Standard Kronecker Tensorial Matrix Product

Consider a matrix $[A]=\left[a_{ij}\right]$ of order (*mxn*) and a matrix $[B]=\left[b_{ij}\right]$ of order (*rxs*). The Kronecker product of the two matrices, denoted $[A]\otimes\left[B\right]$ is defined as the partitioned matrix

$$[A]\otimes[B]=\begin{bmatrix} a_{11}[B] & a_{12}[B] & \dots & a_{1n}[B] \\ a_{21}[B] & a_{22}[B] & \dots & a_{2n}[B] \\ \vdots & \vdots & & \vdots \\ a_{m1}[B] & a_{m2}[B] & \cdots & a_{mn}[B] \end{bmatrix} \tag{A.1}$$

$[A]\otimes\left[B\right]$ is of order (*mrxns*). It has *mn* blocks; the *(i, j)th* block is the matrix $a_{ij}[B]$ of order (*rxs*).

B. The Mueller Matrix and The Kennaugh Matrix

The Mueller Matrix

For the purely coherent case, the Mueller matrix $[M]$ can formally be related to the coherent Jones scattering matrix $[T]$ as

$$[M]=[1 \quad 1 \quad 1 \quad -1][A]^{T-1}([T]\otimes[T]^{*})[A]^{-1}=[A]([T]\otimes[T]^{*})[A]^{-1} \qquad \text{(B.1)}$$

with the 4x4 expansion matrix $[A]$ given by:

$$[A]=\begin{bmatrix} 1 & 0 & 0 & 1 \\ 1 & 0 & 0 & -1 \\ 0 & 1 & 1 & 0 \\ 0 & j & -j & 0 \end{bmatrix}$$

so that the elements M_{ij} of $[M]$ are:

$$\begin{aligned}
M_{11} &= \tfrac{1}{2}\left(|T_{xx}|^2+|T_{xy}|^2+|T_{yx}|^2+|T_{yy}|^2\right) \\
M_{12} &= \tfrac{1}{2}\left(|T_{xx}|^2-|T_{xy}|^2+|T_{yx}|^2-|T_{yy}|^2\right) \\
M_{13} &= \mathrm{Re}\left(T_{xx}T_{xy}^{*}+T_{yx}T_{yy}^{*}\right) \\
M_{14} &= \mathrm{Im}\left(T_{xx}T_{xy}^{*}+T_{yx}T_{yy}^{*}\right) \\
M_{21} &= \tfrac{1}{2}\left(|T_{xx}|^2+|T_{xy}|^2-|T_{yx}|^2-|T_{yy}|^2\right) \\
M_{22} &= \tfrac{1}{2}\left(|T_{xx}|^2-|T_{xy}|^2-|T_{yx}|^2+|T_{yy}|^2\right) \\
M_{23} &= \mathrm{Re}\left(T_{xx}T_{xy}^{*}-T_{yx}T_{yy}^{*}\right) \\
M_{24} &= \mathrm{Im}\left(T_{xx}T_{xy}^{*}-T_{yx}T_{yy}^{*}\right)
\end{aligned}
\qquad
\begin{aligned}
M_{31} &= \mathrm{Re}\left(T_{xx}T_{yx}^{*}+T_{xy}T_{yy}^{*}\right) \\
M_{32} &= \mathrm{Re}\left(T_{xx}T_{yx}^{*}-T_{xy}T_{yy}^{*}\right) \\
M_{33} &= \mathrm{Re}\left(T_{xx}T_{yy}^{*}+T_{xy}T_{yx}^{*}\right) \\
M_{34} &= \mathrm{Im}\left(T_{xx}T_{yy}^{*}-T_{xy}T_{yx}^{*}\right) \\
M_{41} &= \mathrm{Im}\left(T_{xx}^{*}T_{yx}+T_{xy}^{*}T_{yy}\right) \\
M_{42} &= \mathrm{Im}\left(T_{xx}^{*}T_{yx}-T_{xy}^{*}T_{yy}\right) \\
M_{43} &= \mathrm{Im}\left(T_{xx}^{*}T_{yy}+T_{xy}^{*}T_{yx}\right) \\
M_{44} &= \mathrm{Re}\left(T_{xx}T_{yy}^{*}-T_{xy}T_{yx}^{*}\right)
\end{aligned}
\qquad \text{(B.2)}$$

If $[T]$ is normal, i.e. $[T][T]^{T*}=[T]^{T*}[T]$, then $[M]$ is also normal, i.e. $[M][M]^{T}=[M]^{T}[M]$

The Kennaugh Matrix

Similarly, for the purely coherent case, $[K]$ can formally be related to the *coherent Sinclair matrix* $[S]$ with

$[A]^{T-1}=\dfrac{1}{2}[A]^{*}$ as

$$[K]=2[A]^{T-1}([S]\otimes[S]^{*})[A]^{-1} \qquad \text{(B.3)}$$

$$
\begin{aligned}
K_{11} &= \tfrac{1}{2}\left(|S_{xx}|^2 + |S_{xy}|^2 + |S_{yx}|^2 + |S_{yy}|^2 \right) & K_{31} &= \mathrm{Re}\left(S_{xx} S_{yx}^* + S_{xy} S_{yy}^* \right) \\
K_{12} &= \tfrac{1}{2}\left(|S_{xx}|^2 - |S_{xy}|^2 + |S_{yx}|^2 - |S_{yy}|^2 \right) & K_{32} &= \mathrm{Re}\left(S_{xx} S_{yx}^* - S_{xy} S_{yy}^* \right) \\
K_{13} &= \mathrm{Re}\left(S_{xx} S_{xy}^* + S_{yx} S_{yy}^* \right) & K_{33} &= \mathrm{Re}\left(S_{xx} S_{yy}^* + S_{xy}^* S_{yx} \right) \\
K_{14} &= \mathrm{Im}\left(S_{xx} S_{xy}^* + S_{yx} S_{yy}^* \right) & K_{34} &= \mathrm{Im}\left(S_{xx} S_{yy}^* + S_{xy}^* S_{yx} \right) \\
K_{21} &= \tfrac{1}{2}\left(|S_{xx}|^2 + |S_{xy}|^2 - |S_{yx}|^2 - |S_{yy}|^2 \right) & K_{41} &= \mathrm{Im}\left(S_{xx} S_{yx}^* + S_{xy} S_{yy}^* \right) \\
K_{22} &= \tfrac{1}{2}\left(|S_{xx}|^2 - |S_{xy}|^2 - |S_{yx}|^2 + |S_{yy}|^2 \right) & K_{42} &= \mathrm{Im}\left(S_{xx} S_{yx}^* - S_{xy} S_{yy}^* \right) \\
K_{23} &= \mathrm{Re}\left(S_{xx} S_{xy}^* - S_{yx} S_{yy}^* \right) & K_{43} &= \mathrm{Im}\left(S_{xx} S_{yy}^* - S_{yx} S_{xy}^* \right) \\
K_{24} &= \mathrm{Im}\left(S_{xx} S_{xy}^* - S_{yx} S_{yy}^* \right) & K_{44} &= -\mathrm{Re}\left(S_{xx} S_{yy}^* - S_{xy} S_{yx}^* \right)
\end{aligned}
\tag{B.4}
$$

If $[S]$ is symmetric, $S_{xy} = S_{yx}$, then $[K]$ is symmetric, $K_{ij} = K_{ji}$, so that for the symmetric case

$$
\begin{aligned}
K_{11} &= \tfrac{1}{2}\left(|S_{xx}|^2 + 2|S_{xy}|^2 + |S_{yy}|^2 \right) = \tfrac{1}{2} Span[S] \\
K_{12} &= 0 \\
K_{13} &= 0 \\
K_{14} &= 0 \\
K_{21} &= 0 \\
K_{22} &= \tfrac{1}{2}\left(|S_{xx}|^2 - 2|S_{xy}|^2 + |S_{yy}|^2 \right) \\
K_{23} &= \mathrm{Re}\left(S_{xx} S_{xy}^* - S_{yx} S_{yy}^* \right) \\
K_{24} &= 0 \\
K_{31} &= 0 \\
K_{32} &= 0 \\
K_{33} &= |S_{xy}|^2 + \mathrm{Re}\left(S_{xx} S_{yy}^* \right) \\
K_{34} &= 0 \\
K_{41} &= 0 \\
K_{42} &= 0 \\
K_{43} &= 0 \\
K_{44} &= |S_{xy}|^2 - \mathrm{Re}\left(S_{xx} S_{yy}^* \right)
\end{aligned}
\tag{B.5}
$$

with

$$
K_{11} = \sum_{i=2}^{4} K_{ii} = \frac{1}{2}\sum_{i=1}^{4} K_{ii} = \frac{1}{2}\sum_{i=1}^{2} \lambda_i([S]^*[S]) = Span[S]
$$